HOW TO SUPPORT A CHAMPION

THE ART OF APPLYING SCIENCE TO THE ELITE ATHLETE

BY
STEVE INGHAM

SIMPLY SAID

HOW TO SUPPORT A CHAMPION: THE ART OF
APPLYING SCIENCE TO THE ELITE ATHLETE

Published in the United Kingdom 2016
by SIMPLY SAID LTD

Ebook ISBN 9780995464360
Paperback ISBN 9780995464353
Version 1.5 (Edgar)

DEDICATION

To my perfect girls: Rachel, Rosie and Lily.

"You must learn from the mistakes of others. You can't possibly live long enough to make them all yourself."

Sam Levenson

CONTENTS

Preface	1
Acknowledgements	9
Chapter 1: Performance Immersion	11
Chapter 2: The Big Goal	40
Chapter 3: How do you Know?	64
Chapter 4: Seven Spinning Plates	86
Chapter 5: Journeying Through Kocs	120
Chapter 6: Beware of your Thoughts...	153
Chapter 7: Tightrope Rollercoaster	185
Chapter 8: All True to Altruism?	211
Chapter 9: How to Support a Champion: Summary	229
Chapter 10: A Final Word from Me	260
Chapter 11: A Final Word from the Athletes and Coaches	263
Works Cited	269
About the Author	274

PREFACE

"He who works with his hands is a labourer.
He who works with his hands and his head is a craftsman.
He who works with his hands and his head and his heart is
an artist."

Francis of Assisi

I often hear myself and others saying: "There aren't any books telling you how to work with elite athletes."

Well, this book is a contribution to invalidating that statement and sentiment. However, it isn't a point-by-point guide of what you need to *do* to be an effective practitioner. Applied practice is too complicated for a simple checklist manual.

Instead, this book will share with you the intensity, the challenges, the complexities, the strains, the insecurities, the regrets, the mistakes and the lucky scrapes, as well as the fierce ambitions, the hopes, the breakthroughs, the sense of purpose, the joys, the fun, the wonder and the grandeur of being an applied practitioner.

I have written this book as a part of my 'call to arms' for us all to do more to develop and celebrate the art of applying knowledge.

Whether you are a sport scientist, botanist, web developer or air traffic controller, you are a practitioner: 'a person engaged in the practice of a profession'. Therefore, I'll take it as a given that you have a good knowledge base. I assume that you will have done the relevant reading and research, and that you have the certificate to prove it.

Throughout my career, I have been struck by the level of curiosity around the subject of why some people are effective while others are not. Why is it that some practitioners are brilliant, while others are mediocre and some are downright awful? Why is it that when you meet some practitioners they drain all your happiness away, while others brighten up your life? Why is it that, when given an interesting project to work on, some practitioners just go off and do their own thing, while others pull together to provide a collective effort?

Why is it that, when presented with a problem or a question, some practitioners become innovative, sparky and creative, while others just try to hit different problems with the same hammer? When presented with an outcome, why do some practitioners point the finger of blame, while others reflect and accept responsibility? Why do some practitioners just mess with people's business, while others help to make life easier? The difference might be explained by the practitioner's personality, but I think the major difference is whether that person has learned and adapted from their own experiences.

If you fail to learn from your experiences, you won't progress in your understanding. If you fail to adapt from your learning, your skills will remain static. If you fail to learn or adapt, your ability to effect change, and your influence on the world around you, will be limited.

All too commonly, the educational system is stuck in teach-recite or research-write up; two-dimensional methods of training. So how will practitioners of the future ever be suitably trained to work if there is so little *do* in their courses from which they can learn and adapt? Even out there in the big, bad world, practitioners are often afraid to confront the brutal facts of their own performance by self-reflecting, or by giving and receiving feedback, so how will existing practitioners ever learn and adapt higher-level abilities?

I have made thousands of mistakes throughout my career, but I consider myself lucky. I work in the unforgiving, unrelenting and performance-focused world of elite, high-performance sport. In that world, if you don't learn and adapt from every instance, encounter, experience, mistake or failure, there is every chance you will be spat out quickly. Pursuing an ambitious goal, such as a World or Olympic medal, is an environment that is completely intolerant of poor practitioner skills.

On the other hand, if you take the opportunity and make the room to self-reflect, develop, hone, iterate, polish, cultivate, rehearse, nurture and refine your skills, words and behaviours, you will be on the road to becoming an artisanal practitioner.

This book shares the pivotal practitioner lessons I have learned throughout my career working with some of the world's greatest athletes. The first six chapters describe moments when I was required to quickly learn and adapt my professional skills to survive, let alone thrive. I have had the sheer and utter privilege of working with more than a thousand athletes and more than a hundred coaches. The first six chapters focus on six different cases, followed by a further two chapters that address two important concepts.

The first chapter describes my work with the legendary rower Sir Steve Redgrave in the close of his career as he headed towards his tumultuous fifth and final Olympic gold medal at the Sydney Olympic Games in 2000. Steve is one of the most intensely focused sportspeople of all time; completely intolerant of second place and second best. Chapter one describes how I had the challenge of making a connection with him and what he taught me along the way.

Chapter two recounts the seemingly impossible challenge Coach Martin McElroy threw down to me to help transform a group of rowers unable to make it to a final, through to becoming Olympic Men's 8+ champions in 2000. The collective spirit of working toward a team goal, where everyone is pulling in the same direction, both literally and metaphorically, was a compelling objective, but it ultimately questioned what drove me.

Chapter three documents my work with Coach Mark Rowland and middle-distance runners Hayley Tullett and Mike East, who were both dissatisfied by the plateau they had hit. They laid down their expectations of the type of support they needed and how I needed to step up to establish an evidence base they could use. Their torrent of critical thinking would forever change the way I worked.

Chapter four details the support I gave to the dynamic duo of Coach Toni Minichiello and heptathlete Dame Jessica Ennis-Hill as she rose from fledgling junior to 2012 Olympic champion, and on to the present day. The complexities of the heptathlon demanded the most versatile, flexible and imaginative applied practice for me to be able to make any practical recommendations. Some of the avenues of possibility we explored and pursued were completely unexpected, but if we hadn't followed our flow of reasoning and decision-making, we would have been stuck at square one.

Chapter five describes my support work with a rival heptathlete, Kelly Sotherton. Normal scientific support involves the outcome of advising others, but what if the athlete thinks you can do more than that? What if he or she wants you to author their training? In this chapter, I detail what can happen when you cross the Rubicon into coaching.

Chapter six describes my work with monumental rowers Sir Matthew Pinsent and James Cracknell in supporting their goal of winning two World titles in as many hours. I tell the story of how an observation turned into a thought, which turned into a conversation that precipitated more than I could have imagined.

Chapters seven and eight are slightly different. Drawing on an array of case studies, I highlight and explain two crucial concepts that need to be understood and grasped in order to succeed in high-performance. In chapter seven, I untangle the balancing act of progressing an athlete, along with the uplifting highs and deflating lows that come with pursuing a challenging goal. In chapter eight, I highlight the foundational philosophy of supporting others, with altruistic behaviours. I address the dichotomous pull of satisfying your own ambitions while serving those of others and draw on the origins of human technology to illustrate that a variety of applied practice approaches are necessary to progress.

I have chosen to use a mix of storytelling, which I have devotedly reproduced from my extensive note-keeping, and reflective observation throughout the book. The hope is that this blend will amplify, illuminate and punctuate the circumstances encountered and lessons learned throughout my career as a practitioner and leader.

The accounts contain a smattering of technical science, but owing to the dearth of material addressing this area, I make no apologies for focusing on the craft skills of supporting, working and developing others. I hope you can soak up the accounts and reflect on how you would have worked in these situations, visualising yourself developing in a similar way.

In chapter nine, I wrap up all the themes – three for each chapter – into one neat summary, providing you with my top tips and suggested further reading or viewing for you to pursue.

The book closes with some final thoughts from myself, the coaches and athletes, accentuating the need for us to cherish, care for and craft the application of knowledge.

I hope you enjoy the book. If, having read this section, you decide not to read any further, please just get out there, learn, adapt and bring your knowledge to life.

Acknowledgements

Huge thanks must go Dr Emma Ross, Dr Kevin Currell, Dr Jamie Pringle, Andy Allford, Rosie Mayes, Professor Edward Winter, and Colin and Ann Clegg, who generously reviewed and provided feedback on the book. Your input is truly appreciated and cherished.

A special thanks to my wife, Rachel, who reviewed the book several times. Thank you for spotting some ridiculous spelling mistakes (Fanny *Blinkers*-Koen!), giving your highest praise ("This is actually quite interesting, quelle surprise!") and putting up with, supporting and championing me through my obsession with getting this book produced.

Thank you to all the main protagonists – the athletes and coaches featured in the book – who have kindly given me their permission to share the events, situations and discussions from which I have learned so much. Thank you to Steve Redgrave, Martin McElroy, Mark Rowland, Hayley Tullett, Mike East, Toni Minichiello, Jessica Ennis-Hill, Kelly Sotherton, Matt Pinsent, James Cracknell and Laura Finucane. You have all been an inspiration to work with.

CHAPTER 1: PERFORMANCE IMMERSION

"I know you won't believe me, but the highest form of human excellence is to question oneself and others."

Socrates

HENLEY-ON-THAMES

It was July 1998. It was probably Tuesday 7th July, but I can't be sure. I know roughly when it was as I had started my new job on 22nd June. I'd missed the first opportunity to meet Steve Redgrave the week before as he had cancelled his appointment for physiological testing with us up at Northwick Park Hospital, Harrow, owing to a minor injury niggle.

On that day, I had the pleasure of meeting his 'other half', Matt Pinsent. I was twenty-four years old at the time and in utter awe of these totemic legends of British and Olympic sport. It wouldn't be an understatement to say that I had rehearsed the encounter at least a thousand times in my head, over and over, day and night, several times per hour. I did a lot of pacing up and down over the summer of 1998 to stave off the stomach-churning nerves.

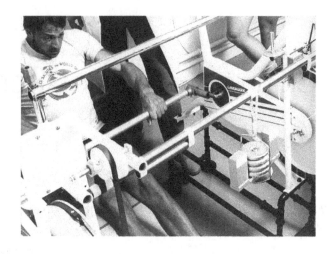

Figure 1. Steve Redgrave performing the 'arm cranking' test before the advent of the rowing ergometer. Unfortunately, the validity of the test was poor. Not only was it limited to testing the arms, even though the arms only contribute around 10% to rowing work, but the cranks were set up in the wrong direction, leading to the rowers dubbing it the 'backing-up test'.

The introduction with Matt went okay. The double doors to the laboratory crashed open, not because of an overly aggressive door opening attack on Matt's part, but because the hinges were called into rapid and unrelenting action under the pressure of the sheer mass of human entering the room. Matt has more than 'enter-the-room presence' he has 'enter-the-room magnitude' and this was enough to give any half-alert human a jolt.

However, just as rehearsed, I strode over, held out my hand for a shake and rolled out my pre-prepared line, "Hi Matt. I'm Steve, your new physiologist." (Don't knock it. It took me ages to come up with that.)

I received the cordially curt response of, "Nice to meet you, Steve. Shall we do this, then?" The intro abruptly over, and with no notable tongue-tangles or brain farts, I jumped into the routine, step-by-step process of taking him through the full rigours of the athlete's regular MOT: the physiological test[i].

My recollection of meeting Matt is stored clearly as a 'flashbulb' memory. It was significant, pronounced and important enough for my brain connections to be super-charged into retaining the images, events and feelings in my version of high-definition.

Fast forward nine days to Henley-on-Thames. I had arrived an hour and a half early, just in case I needed to divert via Wales, and had parked my Citroen Saxo up in the car park at the back of the Leander Club; one of the oldest rowing clubs in the world. With my special Olympic-themed apparel suitably adorned, just so anyone going about their daily business would know that I was 'with the Olympics', I made my way to the boathouse front to meet up with the team again.

The rowers tended to arrive en masse at 7.59am, in plenty of time for the 8.00am session! Sure enough, the giants soon arrived. After several re-greetings of rowers from the week before, I spotted a moment to make my big introduction.

Steve had broken away from the chit-chat, presumably to make his way over to the boat, and this seemed like the ideal moment to make my move. First, I did the walk over ('Excellent, no trips'), then I put my hand out ('All going well, I had managed to lift my hand out in front of me. Superb stuff!') and then came my tried and trusted introductory line (which I hadn't managed to refine

---

[i] *The slave ship rowing scene from the film* Ben Hur *should give you the general idea (youtube.com/watch?v=ax7wcShvrus).*

further from the Matt Pinsent intro): "Hi Steve ('This is going really well. We have the same name, we're going to get along just fine.'). I'm Steve, your new physiologist." ('And relax, this is basically over. All the words were said in the right order. Pat on the back time!')

If meeting Matt had been a flashbulb moment, this was a simultaneously synchronised, paparazzi-style cacophony of strobe lighting, police helicopter spotlights, heaven's calling, aliens are landing, New Year fireworks climax, bright-lighting extravaganza, all while the sun was being fully eclipsed by Steve's massive deltoid frame!

His response was plain and simple, and should have been anticipated, but it wasn't.

"Hi. Are you going to make me go faster?" he fired back.

In a singular instance, Steve sent an Exocet missile to my brain from point-blank range, taking everything I had ever 'learned' in science and either knocking it out or turning it upside down.

The flurry of thoughts going through my mind in that millisecond of neuroprocessing was at storm level of confusion.

'Where should I start?' I asked myself. 'Maybe I should mention my interest in breathing mechanics. No, this chap doesn't look like he lacks for lungs! How about my interest in muscle soreness and growth? No, this man-mountain doesn't look like he lacks hypertrophy skills. And while I stand at a moderate 174cm, I probably shouldn't mention growth to this man towering over me at 22cm in excess of my crown. What about my thesis on overtraining? Yes, that could be a good place to start, but it would be a bit negative and I don't know if he has ever overtrained. Well, it's a better option than any of the others, so I could start there…

'Oh, hang on a minute. Who was one of the most referenced experts in the area? Dr Richard Budgett! Redgrave won his first gold medal with him in 1984[ii]. All right. If not this, that or the other, where do I start?

'Hang on a minute. What had I actually learned during my course? What had I learned that could be useful now in this moment, under this spotlight, while the helicopters are still hovering overhead? What fact or nugget have I been taught that would enable me to kick-start this relationship?'

I couldn't remember a class on prioritisation! In fact, I couldn't remember a class on recalling information while the adrenaline was bathing my heart and driving me to flee!

My knee-jerk reaction, albeit privately in my head, was to start with knowledge, frantically searching for something useful I could state. My mistake was to think that facts and knowledge can be traded for recognition and status, and can act as a pass into the world of these great athletes. Somehow, I had left my undergraduate studies in a state of 'knowing'. All I could think in the summer of 1996 was, 'I know stuff.' I also knew that I knew stuff, which most certainly exuded as narcissism.

'Look at me, people in the street. Look at me with my degree. Would you like to know some facts? Perhaps you'd like to know my degree grades? I did especially well in my third-year Environmental Physiology module, you know.'

I was almost certainly suffering from the Dunning-Kruger effect, where someone fails to assess their competence appropriately[1]. In the early years of experience, confidence is disproportionately high, the devilish irony being that it is one's very incompetence that

---

[ii] Budgett was the one in floods of tears. He heads up medicine for the International Olympic Committee and is one of the nicest men on Planet Earth.

steals away the capability to make a fair and reasoned judgment of one's realistic competence!

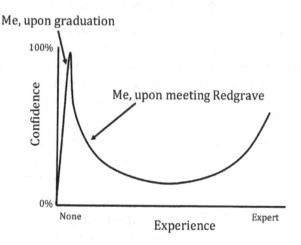

Figure 2. The mismatch between confidence and experience from, 'Unskilled and Unaware of It: How Difficulties in Recognizing One's Own Incompetence Lead to Inflated Self-Assessment [1].

For scientists, the Dunning-Kruger effect is demonstrated by the tendency to turn to published work. To many, this is sacrosanct. To some, it is everything; if it's not published, it doesn't exist. In the clear majority of cases, scientists quote articles as a symbol of their competence, handing them out like greeting cards at Christmas or like food to the needy. I have worked with more than a thousand athletes and I can think of only one who actively wanted to see the literature. Therefore, it wouldn't be unreasonable for me to say that in 99.99% of cases, this doesn't work with athletes; not elite athletes, club-level athletes, or anyone in the middle.

Standing there with Steve on that day in July 1998, I was certainly in the left-hand area of the Dunning-Kruger model. Fortunately, I wasn't ascending the confidence line. I was descending steeply into the pit of appreciating that I might have absolutely nothing to offer.

PRE-HENLEY-ON-THAMES

There were several factors that positioned me on the downward slope of the Dunning-Kruger model. First, while I hadn't taken a 'How to answer Redgrave' class, I was lucky enough to have been taught by some true legends of sports science: Peter Keen, Professor Jo Doust and Dr Steve Bull. They were all fantastic scientists, all compelling communicators, and had had been weathered out in the field, applying their science and able to share their lessons. Without this grounding, I would almost certainly not have been able to visualise a similar career path for myself.

Second, I had clocked up my own experience of working with recreational and regional-level athletes. Some were quite good, which added to the experience perspective. I had done so under my own steam while I was studying, and at a regional sport science unit in Worcester.

At the heart of my experience, I had looked into the whites of the eyes of many athletes. I had seen their reactions to my way of working and heard the sighs of discontent when my manner wasn't quite right. I had seen the jaws jut in, the eyes flicker from side to side and the body language close up when I had suggested something alien to them. As blunt feedback goes, the silence of not hearing back from an athlete when it's in their best interest to get scientific support is the loudest shout of rejection. The silence tells you they would rather go without than work with you.

Third, I had reflected a lot on my experiences. This I owe to Ailsa Niven (née Anderson), who was studying reflective practice while I was at Worcester. She asked me (actually she insisted, in a cheekily polite way) to undertake a formal reflective process (see example overleaf[2]) each time I interacted with athletes and coaches.

At times, it felt like I was reflecting on everything: 'How did the session go with that triathlete today?', 'How did that gas analyser calibration go?' However, without it I know I wouldn't have been in a position to take the Redgrave Exocet. In each instance, the loop of asking sound, reflective questions meant that when I had hit the sweet spot of reflection in 'analysis' (see inset below), I simply couldn't avoid confronting the big issues of how it had gone. So, when it came to my own performance, my 'scientific skills' were in competition for my attention with 'my rapport with the athlete'.

Critically, my previous experience positioned my respect at the door of athlete and coach experience and understanding. What are the athletes and coaches actually doing (because I need to know)? How does the athlete and coach think they are getting on (because I need to know)? As scientists, we are trained to know, albeit based on statistical probabilities, cross-study perspectives and reviews, and when it comes to communicating with other people, this process cultivates you to recite facts, definitions and findings.

As an applied scientist, you should be motivated to influence the world around you. Simply reciting information is negligent of the context, the environment and the individual circumstance. Therefore, an applied scientist aiming to improve performance should be motivated to ask questions and glean knowledge.

Steps of reflective practice:

- **Description:** 'What happened?' (judgement-free)

- **Feelings:** 'What were your reactions and feelings?'

- **Evaluation:** 'What was good or bad about the experience?'

- **Analysis:** 'What sense can you make of the situation?', 'Bring in ideas from outside the experience to help you', 'What was really going on?', 'Were other people's experiences similar or different in important ways?'

- **Conclusions (general):** 'What can be concluded, in a general sense, from these experiences and the analyses you have undertaken?'

- **Conclusions (specific):** 'What can be concluded about your own specific and unique personal situation or way of working?'

- **Personal action plans:** 'What are you going to do differently in this type of situation next time?'

BOMC

Back to the hinterland between Redgrave's response and mine. Although I wasn't prepared, I was primed. By chance and/or by good grounding, I had done some homework. I had been given the responsibility of working on the regional rowing programme up in Worcester and, in a moment of frustrated incompetence, I had contacted the National Programme down at the British Olympic Medical Centre (BOMC: the forebear of the British sports institute systems) and asked to visit. I wanted to get the lowdown on how these tests worked and how they were used, all to enable me to look like I knew what I was doing with the West Midlands rowing team. Therefore, I had a relationship with the physiologists in situ.

When the advert came out for a physiologist at the British Olympic Medical Centre and I got the call for an interview, I phoned one of the staff and was surprised and confused by the advice I received. My initial question was, "What's it like working with these legends?"

My colleague responded thus: "Well, first what will happen is you'll start working with these guys, be surprised by what they are doing, try to influence it and then get frustrated after eighteen months and want to leave. I'm going off to do a PhD, to do some real science. That's what everyone else has done."

Cue a bemused frown from me.

With some light questioning around what was creating this frustration, I was further told in an exasperated tone, "They're doing all this mileage, all these low-intensity miles. Hours and hours of plodding back and forth, with virtually no intensity and no threshold/tempo work[iii]. The event is only six minutes long and the training they're doing is irrelevant, but they just won't listen. I've tried to tell them they should be doing some threshold work; some high-intensity intervals to develop race-specific

---

[iii] *The exercise intensity where you are clipping along at an intensity at which you can only just hold a conversation. It's called your 'threshold pace', as it marks the level beyond which you are leaning on your anaerobic energy systems much more, so if you push any harder you won't be able to keep it up for long.*

fitness. I've shown them the papers and explained it to them, but they just won't listen."

I was getting the impression that this was going to be a tough gig, so I asked why they were doing what they were doing. "Well Jürgen [Gröbler, coach of the GB rowing team] is stuck in the 1970s. They had considerable success using this low-intensity approach then, when no one else was using a structured training approach. We know so much more now. As for his willingness to take on new ideas, well, forget it."

'Blimey, he doesn't think much of the most successful coach in Olympic history,' I thought! I asked how the athletes felt.

"They just go along with it. They do what they're told because they want to get into the top boat."

I later found out that if you dig much deeper, Jürgen's approach, the athlete's interpretation and the rationale for this type of training is entirely sound and not only produces great training gains, but builds stable, predictable performance abilities. What more do you need for developing high performance in rowing?

But during this telephone call to my colleague, I was stunned. Not a bone in my body wanted to go down there and tell them what they needed to be doing, as he suggested. My professional standing as a sports physiologist, a support scientist, let alone with sports specific expertise in rowing, was flimsy[iv].

What was I going to say? "Have you heard about my work with the West Midlands rowing team? Last year they were sixth at nationals and this year they got a bronze. How d'you fancy a piece of that pie?" How on earth would

---

[iv] *Some would say little has changed over the years.*

that go down with Redgrave and his (then) four gold medals?

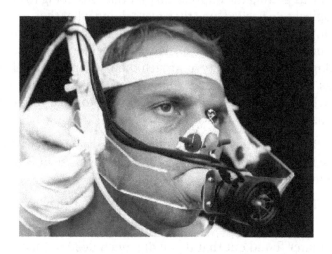

Figure 3. Steve Redgrave undertaking a physiological test at the British Olympic Medical Centre, circa 1998.

In a perfect world, I would have loved Steve to have met my, "I'm Steve, your new physiologist" with a, "That's great. Nice name, by the way. Great to have you on board. I'm really looking forward to hearing all about what you know and to perhaps have a read of your undergraduate dissertation abstract. Hey, why don't you come and have a look at my boat? It's really shiny. I'll tell you what, why don't you wear some of my medals and we can skip there together?!"

This was never going to happen. All that I had heard from the outgoing physiologist was bewildering because there appeared to be no credence given to why he was doing what he was doing. There must have been a reason for the athletes' choices. They were actively choosing to use a low-intensity training approach, but also actively deselecting other methods; in this case the threshold/high-intensity training.

SHOOT ME

From everything I could see, these were not reluctant champions. You only have to delve a little into Steve Redgrave's media quotes to learn of the intense all-or-nothing nature he lived by:

"If anyone sees me going anywhere near a boat again they have my permission to shoot me."

This statement was growled in the humid heat of Lake Lanier at the 1996 Atlanta Olympic Games, shortly after having secured his fourth gold medal, Matt Pinsent's second gold, and Team GB's only gold medal of the Games. Steve had ploughed everything into winning in 1996, then suddenly became very focused on never rowing again.

Gold Fever, the BBC documentary about the coxless four's preparation for the Sydney Games further shows that this is not a man who messes around. In the documentary, Steve is shown to push himself into unconsciousness, almost certainly due to a hypoglycaemic (low-glucose) state brought on by his diabetes, but also as a result of the 2,000m all-out ergo test that pushed him over the edge. Steve would have known that he was in trouble all the way through the test, but his refusal to stop perfectly highlights his willingness to drive his body beyond its limits to make it ready for competition.

I just could not visualise a scenario in which any attempt to flex my authority, knowledge or standing would cut it. I wasn't totally convinced that the ever-so-wise approach of asking questions would be the best one either.

"Hi. Are you going to make me go faster?"

"Well, perhaps you could tell me what you're currently doing?"

No, no, no! That would sound like I had a clipboard with a fifty-point needs analysis to administer. Could Redgrave's question have been rhetorical? More than likely not. So, if this is not a request for information and if it's not rhetorical, it might be a search for an understanding of my capabilities.

No, I don't think so. It hinted at a need for some sort of assurance. Given what I had heard on the phone about what it was like working with these elite athletes, I already knew that a line of previous physiologists had tried to change their training. They had tried and failed to tell them what to do, then got frustrated and left.

So that's what Steve had experienced: numerous scientists rocking up and telling him what to do. I didn't want this to be me. I didn't want to hit the same brick wall. I wanted to build meaningful relationships with these people. I couldn't conceive of my career starting and then ending just as quickly.

The unsettled sleep, the preparation of my opening line, the pacing, the nerves, the flurry of thought and the intimidation I felt were all in proportion to the reverence I felt for these colossal athletes; these monumental legends; these gargantuan gods of British sport. I am, at the heart of it, a sports fan. I would find myself sitting at home as a six-year-old watching the highlights of the Moscow Olympics, then even more of the Sarajevo Games aged ten, more again of Los Angeles aged fourteen, then all of Calgary, Seoul, and each one since as the years flashed by.

The Barcelona Games in 1992 was a tipping point. Having watched it all, and primed by my studies of human biology, I was inspired to wonder, 'How do they do what they do? Why can they run so fast, jump so high and be so strong when I clearly can't?'

So I had watched all of Steve's victories, all his golden moments. I had seen all his interviews and collected the collectable magazines, so I knew as much as the lay media would allow me to know. I knew he was busting his way to his wins, that he was relentless, and that he was intense and intolerant of failure. I knew that working with him and the team around him would not be a walk in the park, but I was massively ambitious to do so and wanted to succeed.

Narrowing it down, my perception was that his response was probably a way of questioning me and my intent.

"Are you going to make me go faster?" could have been interpreted as, "Are you going to get in my way?" This certainly put me on the back foot. In fact, it put me dangerously close to losing control over my bodily functions. Fortunately, the development and training I had received up to that point in my career came to the rescue.

This is how I responded:

"I don't know whether I'm going to make you go faster, but I'm really keen to find out what makes you so successful, and if I find anything else out along the way, I'll let you know."

Then followed the shallow breathing, heightened eyebrows and eyes focusing on every bit of facial and body language going, ready to hear whether I had messed up or not.

"Hhmm." Pause. "We'll see," he said with an unimpressed flick of the head.

My latest interpretation could now be rephrased as, "Are you going to be annoying as I try to summit this fifth and tumultuously final mountain? Because if you are, it's going to weigh me down and I won't let that happen. You'll be chewed up and spat out! So decide now if you're going to annoy me and hinder my climb, and if there's a chance you'll be a burden to me, stand back in the shadows and just do your job."

The lack of a pause, flinch or concession in Steve's step from "Hi" to "Are you going to make me go faster?" indicated this intensity, this intolerance, this lack of patience with those who had tried to work with him without journeying with him.

Journeying

Working with another person in a simply transactional format might as well take place from the other end of a telephone:

"Should I drink 'A' or 'B'?"

"I recommend you drink 'A' because it gets into your body faster."

"Thanks. Bye."

On the other hand, journeying with an athlete is about making the travel easier at every turn. It's about being a good partner and a good support. We have all been on physical journeys that take an age, perhaps in a hot, sweaty rail carriage, on a bus with an annoyingly hyper group of school kids, or in a car when we're simply longing to get home.

Equally, we have all been on journeys that should feel long, but somehow the time flies by. Perhaps we're distracted by the stunning landscape of the Highlands or the Rockies, or maybe we're enjoying charming company.

The same is true when it comes to supporting an athlete. Are you going to make this experience easier, helping to propel them at a faster rate and increasing the probability of them reaching their summit? Or are you going to burden them, slow them down and hinder their ascent?

Working with talent requires you to be a journey partner; to walk in the footsteps of the athlete and coach; to be willing to put the hours in and happy to carry the bags; to wait patiently to be asked to contribute. Then, and only then, do you get to continue that journey. It's important not to forget that nurturing someone else's talent every day, in a situation where you and they are testing the possibilities, solving problems, seeing progress and achieving (or at least trying to achieve) their goals, is very cool!

The first level of acceptance is, "Do I like you as a person?" Or perhaps, "I might dislike you, but are you going to be a negative influence or a distraction from my performance?"

Second base, maybe, just maybe, seemed to be, "I might ask your opinion," but that was a long stretch from my introductory experience down by the river. I perceived that my answer had at least avoided giving annoyance, and I was chuffed to bits to know that.

The subsequent months required me to get my head down and do the basics of my job faultlessly, with no fireworks and no surprises. I was to be on time, take bodily fluid samples rapidly and efficiently, care for my machines and have backup plans that might take days to create without always being needed. Often, they were there just in case.

I was expecting to begin travelling with the team and, given the time of year, that was likely to start soon. Within a matter of weeks of meeting Redgrave and this team of giants, I was invited to Austria for an altitude training camp. The team flew out while I drove a white Renault Trafic van, complete with all the testing kit and with a top speed of 50mph, to the Tirol region, where I would encounter thirty-four hairpin bends on the route to the Silvretta training camp.

In preparation for the trip, I had asked one of my new colleagues what it was like up there. Again, the response was baffling: "Silvretta is the arse end of nowhere. It's a miserable place, with mainly mist and cows for company. Bring something to do because it's the most boring place on earth."

The truth was light years along the other end of the spectrum. Yes, there is mist (cloud actually), and yes there are cows (fine-looking ones), but Silvretta is a beautifully inspiring, majestic location, set in the Austrian Alps and cradling a glacial lake formed by an impressive dam. In the shadow of these breathtaking mountains sit two wooden scout huts, where we were to reside.

HOT BUNS

Acceptance was a slow burn. Steve and the others would test me here and there, quizzing me on the previous day's results just to see if I remembered.

In sports physiology, a common measure is to test for the concentration of the metabolite lactate. Anyone familiar with this term[v] knows that it has become, mainly through the continued campaign of sports physiologists,

---

[v] Not to be confused with the lactation of breast milk post-partum!

synonymous with that dull sensitisation of nerve endings telling you that it burns in the muscle when you push your muscles a little beyond the level they are happy with.

Some sports use blood lactate concentrations as a routine measure of whether an athlete is training at the level prescribed by the coach, though in today's sophisticated support systems the use of this measure is dwarfed by the individualised support, which places lactate and all other measures as tools to answer questions rather than as an end in themselves. It requires a small (20-30 microlitre) capillary sample of blood taken from a pinprick lance to the earlobe[vi].

This should take about thirty seconds to obtain, but it requires some skill to take it and can be tricky if it's cold or the intensity is high, both of which cause blood to be diverted from the periphery. But it is possible to get it no matter what, given plenty of practice.

Some skills look deceptively easy but can be a bugger to master, and capillary sampling is one. Speedy blood sampling is a basic skill that instantly makes you look incompetent if you don't have it. If you are taking ten to fifteen seconds to get your sample you'll be fine, but if it takes forty to sixty, and you're squeezing away at an earlobe waiting for some blood to arrive, you're likely to be met with some agitation and possibly even some infuriation from the owner.

This would be equivalent to a strength and conditioning coach being slow in assisting a lift when 'spotting', a biomechanist forgetting to align the speed gun, a psychologist not listening or a physiotherapist working on the wrong limb. These skills are the absolute basics from which applied brilliance can spring, but when they're absent, incompetence is presumed.

[vi] *A line in body piercing is always available if things don't turn out well.*

Athletes get tired a lot. Rowers flog their bodies twenty-four hours each week, cyclists thirty-five hours each week, and runners a little less at about fifteen to twenty, as they have impact to deal with. Each sport varies due to the specific demands of the movement. When tiredness, excessive fatigue and overtraining kick in – and they almost always will when you train this much – mood takes a nosedive and people get fractious at the smallest of things. Redgrave wasn't immune to this and, added to the heightened intensity, the fact that he was getting older and that he had diabetes, this led some people to question his place in the top boat. He would often recoil when the blood sampling took longer than usual.

"What are you doing back there? Get on with it," and the old faithful, "Is this going to make me go faster?" In response to this one cold, damp, misty morning in the Silvretta-Haus garage at altitude (glamorous, eh?), I bristled in a mildly hoity way, "Yes, I think it is!" Like a good boy, I reflected on how well that went and came to a swift and resounding conclusion: not at all well.

The necessity in my early career and never to be forgotten since has been the importance of doing the basics brilliantly. In the early stages of my work, this was like a swan gliding across water, where the work would go unseen, paddling away underwater to make the basics and the complex appear simple. Later on, as these skills become routine, doing the basics will, of course, become easier and possibly even mundane. But there are often few shortcuts to quality assurance, and when it's not there the whole stack of cards will come tumbling down.

Two events triggered a step change in my relationship with the rowing team, and particularly with Steve. Both occurred in the space of our first two weeks at Silvretta. The first was a bit of luck. Back in the days of pigeon post, I would write a letter or two each week to my future wife,

Rachel. I would pop up to the Silvretta shop, buy an Austrian stamp for England and pop it in the post to be collected at 3.30pm every other day.

Being a cheeky monkey, I would put a pair of brackets between her first and surname and include a saucy name. Something like 'Sweet Cheeks' to give you the gist! One evening I was running a little late for dinner, with a host of analyses to process in my little makeshift lab in the scout huts. I arrived for dinner; which was hosted in the hotel's main function room, ideal for hosting a dinosaur-sized, forty-person buffet. Just by the food table to the left of the room was a little space where letters to members of the GB rowing team would be left. As I entered the room, slightly flustered and hungry, I was greeted with a gathering roar of "Wahey, hot buns!", "Show us your buns" and "Smoking!"

One eyebrow went up as I looked behind me. Someone kindly pointed me in the direction of the sole letter on the table. In familiar handwriting, it read 'Steve (HOT BUNS) Ingham, Great Britain Rowing Team'. Now with both eyebrows raised, there was a rush of cringing embarrassment as I slowly fathomed that my brassy shenanigans had been fairly retributed by the missus.

As I took a steady intake of breath and smirked awkwardly, I turned to face the wave upon wave of banter. I was to be affectionately known as 'Hot Buns' or 'HB' by the rowing team for as long as I worked with them, and many still use the nickname to this day.

The 'HB' incident wasn't something I would have chosen to happen. Admittedly, though, it gave me a moniker, an identity. It showed I could take a bit of ribbing and, as such, integrated me into the team at a much faster rate. A purely accidental situation that could never have been manufactured gave me a thankful initiation into the team. It showed me that, in any given moment, in any given interpersonal situation, you need to give of yourself. You need to get involved, share who you are and let down the façade so others can connect with you.

This is why team-building exercises so often start with finding out about yourself and then building to finding out about each other. Self-awareness is an observable cornerstone of high-performing people in the elite sports industry. Those who can seemingly influence others so effortlessly can only really do so if they have an acute sense of self.

THE ROCK

Figure 4. Steve Redgrave looking over the results of the morning's 'baseline' testing. The expression says it all: "Just because I'm looking, doesn't mean I'm all that interested!"

The second moment that changed my relationship with Steve originated purely from spending time with him. Steve seemed to keep me, the physiological support service and the testing results at arms-length. He gave off an air of the uninterested.

One thing I noticed in the first few weeks working at altitude, was that, once I had posted the results of the morning's waking measurements and early morning results on the corridor walls of the accommodation, the first person to creak out of his room to look over the results would usually be Steve. He would spend a good ten minutes combing over the results. That was more than enough time to soak up his own data, so I presumed it was also enough time to see how everyone else was getting on. Who was struggling? Who was finding it easy?

After a week or so of me buzzing to and fro between my room, the laboratory and the kitchen, nodding to the big man with the occasional, "Alright Steve," my head was full of questions about what he was thinking and interested in. I wondered whether he had any questions.

After several surges of 'I'm gonna ask him' followed by 'I don't think you should' or ' I don't think I will this time, maybe tomorrow,' I broke my cyclic suppressed inner voice and ventured, "Kettle's on. Do you want a cup of tea?" This is the only true way for a Brit to break the ice with another Brit!

To my self-satisfied surprise, he turned his head, looked at me and said, "Yeah, go on then."

'Blimey!' I thought.

And so followed one of the many conspicuous inner-monologue statements that summer: 'I'm making Steve Redgrave a cup of tea!', 'I'm testing Steve Redgrave's urine right now', I have Greg Searle's sweat all over my arm', 'Matt Pinsent just stood on my toe. How cool! That's Olympic-level pain!'

"How about outside?" I ventured.

"Yeah, okay."

So we each sat on a rock outside the scout huts and, in the company of the dam wall and the Silvretta Alps, I learned more about what it takes to be an elite athlete, a champion and the relevance of science than I ever had done before or ever have since.

I asked him how he was getting on and how the training was going.

"Not so bad," he replied. "I'm not quite where I'd like to be, but I'm strong enough. We should win at the [Cologne] World Championships."

I interpreted this as him having taken a bit of a dip, and I could see this in some of the physiological testing scores and the training splits he was holding. He wasn't at the top of the pack. What I didn't know was how all those abilities hung together into a 2,000m performance on the start line in a boat against other men. I was captivated to hear Steve let down his guard in this way. He wasn't, after all, invincible.

I then asked him what he thought about 'all of this science business'.

To start with, he gave an insightful response: "The information can be useful. At times when I'm tired, the data might show that I'm okay, so it makes me wonder if I should just crack on. Then, when I'm feeling okay, the results might show that I'm struggling. It gives you a bit more information and allows you to weigh things up a bit."

I took it a step further by asking what he thought about the way the science is delivered.

Steve moved through the gears. "Generally, it pisses me off," he shot. "You lot tend to come in and tell me what I should be doing and how I should be doing it, but you don't realise I've tried it all before. For 1984 [the Los

Angeles Games], our coach, Mike Spracklen, got us doing a lot of high-intensity interval training and we won. For 1988, he got us to do the same, but in added some weight training and we won. When Jürgen came in [in 1992], he changed it all around and we did a lot of low-intensity mileage with some weights, and we won. For 1996, we still did a lot of mileage but with more weights. We still won.

"Scientists come and tell us we should have this special system, but we know it's based on some half-bent study on students. It probably doesn't matter a great deal about how we train; we just need to do a lot of it and with plenty of variety.

"The thing that grates more than anything is that scientists seem to think they know best and have no respect for what we've learned over the years. In over ten years, you're the first sports scientist to ask me what I think. In my experience, scientists think they know it all, when, compared to what we know about high performance, scientists know nothing."

It was bombastic stuff from Steve as he vented deeply held feelings and cutting observations.

I was mightily relieved to have asked the question and was rapidly trying to think of another. The third revelation came from asking him how he felt about going for his fifth gold medal.

He told me, "I know it won't feel like success for me. If I get bronze, I will have failed. If I get silver, I will have failed. If I win, it's expected. So it's just normal or failure. I just can't win."

Taken aback, I asked about the allure of the Olympics being in held Sydney.

He conceded a little: "Yes I expect it will be a great Games. That was part of the motive for me carrying on. But for me it will only be repeating what I've done before and, given the level of expectation, especially as we were the only ones to win in Atlanta, it will only be relief for me.

"I'm not expecting jubilation. That's what comes when you're not expecting the win. You see it all the time: someone wasn't expecting it and they pop up and take the opportunity; it's a surprise. That won't happen to me, because I've got too much pressure to win and that rarely happens in rowing. It's a 'form' sport. You should get the result you deserve. If I win, that's what is expected."

My initial reaction was just how extraordinarily hard on himself he was. But this is Redgrave; incredibly intense, unrelenting, unforgiving and intolerant of anything other than achieving gold in Sydney, regardless of the emotion. For me to achieve acceptance with this sporting giant, for me to be allowed to journey with him and the team at this point in their climb, I needed to provide them with assurance.

Steve had called for it. "Are you going to make me go faster?" was now officially translated as, "I can accept you if you're not annoying."

The discussion on the rock was a three-hour tutorial on what it's like to not only be elite, but the elite of the elite. In so doing, he called for trust.

This now sounded very clear: "You can journey with us, but you must balance your messages so it complements what we're doing. Respect how hard our training is and how much we must push ourselves. Recognise what, how and why we do what we do. Try to learn from what we already know, because we know you won't have seen anything like this before. Understand we are under incredible pressure to perform under the brightest spotlight and that changes the game from being a nice sporting experience to an intense exposure of our souls."

This was the no-nonsense, brutal performance focus Redgrave shared with me. There is little I can say that will add to what has already been said about Steve Redgrave. In the joyous moment the coxless-four's boat crossed the finish line in Penrith, Britain finally woke up to the fact that it had a bona fide hero. His five gold medals at five successive Games placed him firmly among the Olympic all-time greats, and for good reason.

When you consider that he was only just making ends meet after his fourth successive gold medal, had battled through numerous injuries, struggled with colitis, developed diabetes and won his last gold at the age of thirty-eight, you will only just realise what an extraordinary person Redgrave is. His attitude, focus, intensity and drive is a high tidemark in recent sporting history. He is a reference point by which all others pursuing a goal can be measured.

Whenever I see a glimmer of the same attributes or attitudes, I call it the 'Redgrave factor'. It hints that somebody has some of the components of what it takes to achieve extraordinary things.

As we entered the Olympic season, Steve began to relax. Once he had got past trials and the World Cup races (with a minor hiccup, more on that in chapter six), which serve as the starter to the main course of the major championships, Steve remained focused when needed, but he warmed up considerably. He was fun, a good laugh and full of banter around the table-tennis competitions (I lost to him in the semis). His confidence became palpable and his performance was back to his best.

When we arrived at the Olympic Association's holding camp in the Gold Coast, pre-Games, the rowing team held a meeting to cover logistics along with team plans and expectations.

Steve gave a team talk. He spoke about his experiences at his first Olympic Games in Los Angeles, where he had been distracted by the Daley Thompson video games in the village and suffered thumb and forearm fatigue, which had affected his rowing. He gave insights from the 1996 Opening Ceremony and recalled the reverence and honour of watching a fragile Muhammad Ali carrying the torch into the stadium.

Figure 5. Steve Redgrave and me at the Hinze Dam, Gold Coast, Australia, a few weeks before the Sydney Olympic Games. (Reproduced with permission from Professor Ron Maughan, copyrighted).

But he also warned that the ceremony would undoubtedly be tiring, encouraging the team to prioritise their performances. He then finished with a rousing pitch for everyone to row to their true potential; to let their competitors be the ones to make the mistakes; and to take the once-in-a-lifetime opportunity that comes with being an Olympian.

I was seated next to Miriam Batten, a member of the Olympic silver medal winning quadruple sculls, who turned to me at the end of the talk and asked, "Are you cold? You've had goosebumps the whole time."

"No," I replied, a little embarrassed, "That was just so inspirational!"

Redgrave demanded trust that summer, and I somehow managed not to break it. Helped along by some experiences that primed my mindset to want to help, support, learn and not recite, quote and instruct, I was ready to handle the confrontational greeting that acted as a gatekeeper to acceptance from the team. I learned that:

- Focusing on doing the basics brilliantly kept unnecessary criticism at bay and showed I could be relied upon.

- Showing a bit of my character ('Hot Buns') creates a connection with others and develops rapport.

- Asking questions unveiled and illuminated a whole world of understanding about the elite athlete. It also showed I was interested in and respected the views of the talented people I was working with.

Chapter 2: The Big Goal

"Life's most persistent and urgent question is, 'What are you doing for others?'"

Martin Luther King Jr.

Testing times

A chap called Ben Hunt-Davis had a slightly different adaptation on Steve Redgrave's enquiry of, "Are you going to make me go faster?"

He and his rowing crew adopted a similar mantra, which explained the attitude and approach of their pursuit: "Will it make the boat go faster?"

He describes this extremely well in his book of the same name[3], which documents the crew's pursuit and the underpinning psychology behind their golden success. It is well worth a read.

Ben was one of the first rowers I got to know quite well. He was on the peripheries of the top boat, certainly one of the strongest physiologically and, at 198cm, one of the biggest.

My routine at training camps was to get up before any of the rowers, normally at 5:30am, and begin to set up, waking up and calibrating my analysers and myself. By 6:00am I would have my first pair of rowers, normally roommates, come in for testing. Generously, they would all bring me a 75ml portion of their morning's urine excretion. Aren't I a lucky boy?!

One rower would sit down and wait a minute while I sampled his (or her) ear for blood and get the analysis chugging away. At the same time, the other rower would strip down to his rowing all-in-one and weigh himself, logging his waking heart rate, how well he had slept and any soreness or tiredness he was feeling. Then he would wait for his blood to be sampled.

In the meantime, I would be juggling away with blood sampling, blood analysis, recording blood scores, urine sampling, urine analysis, recording urine scores, resetting heart rate monitors, handing out sports drinks as they came in, changing latex gloves between each person, cleaning the machines between each sample, fielding questions and giving out advice all within 150 seconds per person while taking care not to breathe my morning breath on anyone, all before the sun rose. This wasn't the place for existential questions, suggesting a sing-song, or, as much as I may have wanted to, to be grumpy!

My duty was to tread a delicate balance of quiet effectiveness, delivered with just the correct amount of reciprocal banter, while also gleaning how folks were coping. The rota of ensuring twenty-five people got tested in less than an hour meant an athlete's morning slot got steadily later by five minutes each day, until you reached the last slot and then, ouch, they were back to 6.00am again.

This, plus the camp fever that gradually crept in as the weeks went by, meant the entrance greeting from the rowers would predictably worsen. I would normally be happy with a gravelly, monotone, "Muuurnin'", with an imperceptible movement of the lips and jaw. The sleepy dust was still in, the hair would be free-form and the morning emissions were a bit muggy, so this was about as good as it got. More usually it would be a bit more Neanderthal-like, something like, "Hhmm", and, at a stretch, an eyebrow flicker.

However, Ben was in a class of his own. Come rain or shine, 6.00am or 6.55am, tired or otherwise, he would appear at the door, filling the frame as they all do and stoop to enter, typically in a sarong of some description (David Beckham must have got the idea from Ben). With shoulders back and his head up, he would sing with the projection of all three tenors, "MOR-NIIINNNG."

No matter your mood, you couldn't help but be feel uplifted by it. That's Ben all over.

THE PROGRAMME

I had already established that the training programme was not a 'back to the drawing board' topic of conversation. They didn't want me messing with what was a tried and tested schedule of training, which was largely unchanged since it had been introduced in 1991. I also discovered that it was ferociously hard. At a camp, there was a modicum of concession to begin with: a gentle walk or a light paddle of the boat just to overcome the travel, altitude or heat. Then it got harder and stayed very hard.

The altitude camp is a 'work' camp, profoundly stressing the body, after which a tapering camp increases in intensity but greatly reduces the total amount of training. If the body is not subjected to too much physical stress, the taper alleviation *should* allow the body to recover and then adapt to a new higher level.

Finding the right level of training to impose on the body is a difficult balance. It has to be hard, and I mean hard; especially for elite athletes with big engines, a big capacity for training and often a long history of doing so. The long-standing training programme in use with the men's heavyweight rowing team was generally built around Steve Redgrave and Matt Pinsent. They were the top dogs and had spent the previous seven years together soaking up this programme.

There is a great story about when they were first training together. Steve was twenty-eight and Matt was twenty, so Steve was the senior athlete with two Olympic gold medals in his locker, while Matt was still at Oxford University. When they first trained together, Matt, who sat in the bow seat, would get 'pulled around'; when one of the rowers dominates the other with their superior strength, which swings the direction of the boat off course. The physiologists who worked with them reported that Steve's blood lactate would be minimal in the 2-4 mM concentration range, whereas Matt's would be very high in the 6-8 mM range, suggesting he was working much harder, and therefore suffering much more, to keep up. They plotted the training changes and saw that Matt quickly adapted, matching and overtaking Steve in the space of a few seasons.

Matt's physiology was incredible. All rowers have a big aerobic capacity, but Matt was the most blessed aerobic athlete I worked with. His lungs were massive, knocking on 9 litres, and his VO_2max, the maximal aerobic capacity, was regularly greater than 7 litres per minute; both comfortably double that of the average man.

The closest in aerobic capacity was Greg Searle, then Steve Redgrave, James Cracknell, Ben Hunt-Davis and so on. Roughly speaking, the higher the aerobic capacity, the higher the capability to absorb training. The training programme bar was set high, so if anyone adapted as Matt did they would excel. In the main, though, three-quarters of the team just hung on for dear life.

About halfway through that first altitude training camp, Ben came and knocked on my door.

"Hot Buns, I wonder if you can help. I'm really struggling. I'm just so tired and I'm not sure I'm coping with the training very well. Is there anything you would recommend?"

My first proper request for advice had just arrived! I sat Ben down and asked several questions about how he was feeling and what he was doing. We pulled a plan together to tighten up on his diet, improve his hydration, tidy up his sleep pattern, including a proper afternoon nap, and to take the edge off some of his training sessions. I also gave him the reassurance that he had a great engine and, of all the rowers at the camp, that he was well-placed to absorb the training and not break down.

Ben was in the men's heavyweight 8+ crew and had been a stalwart of the boat for many years. The 8+ boat typically hovered between seventh and fourth; out of the running for medals.

This level was generally accepted as about right, primarily because the crew reflected the bottom rung of the 'world class programme' of funded athletes. If you were to look at the range of physiology of the men's 8+ crew, you would not have said they were ripe for winning. On the surface, they just didn't seem to have the horsepower, or quite possibly the hardware, to compete at the top end. In 1998, they came seventh at the World Championships.

Ben's question hinted at more to come from him and from the crew.

Bottom line

As part of a routine physiological testing session, it was my job to collate the test results and sit down with the rower, and possibly the coach, to take them through a rather dusty presentation of results.

In November 1998, I sat down with one of the rowers after the test, along with Martin (Macca) McElroy, the men's 8+ coach. I launched into my description of the data, starting with the body mass and body composition (an estimate of the fat and lean proportions).

By the time I reached the main body of the test feedback, Martin interjected with, "Look, Stevie, what's the bottom line here?"

Unable to wrestle away from my formulaic reporting of a long list of numbers, where the main features were whether the results had gone up or down, what they had been like this time last year and whether we should interpret that as a genuine change or not (known as error of the measurement, if you are interested), I carried on with my account.

Macca sat there, a little nonplussed by my decision to ignore his question. The feedback session ended and he took me aside to make himself a little clearer.

"So what?" he challenged. "I know physiology is important. We all know that. But what do I need to do differently, or should I stick to the main training plan? I've got a good bunch of athletes and I want them to have a good chance of success this year, so I need to know if there's anything I need to do differently."

I sat back and took a moment to think. Macca appeared to be asking for a prioritisation of what was important. Perhaps, most engagingly, he was asking for it in a way that signalled a need for assistance.

I zoomed out my perspective: "Essentially, your athlete here hasn't improved a great deal over the last few years."

Martin coaxed for more: "So has he hit a ceiling or is the training misfiring?"

I was being asked for a judgement on the entire state of play with limited but sufficient information.

"I don't think he's hit his ceiling. He should be continuing to improve at his age and stage of training," I concluded.

While I put forward my proposition, I could feel that I was stepping beyond the norm. In previous jobs and in my initial steps into the Olympic arena, I had been cautious about venturing too far into the 'recommendations quagmire' that Redgrave had so clearly warned against. I had simply been putting forward my measurements, explaining what each of them was, together with the consequences of them going up or down, and signposting people to the choices and steps available. My work was pretty much completed at arm's length, as I wanted to avoid making mistakes or appearing to be arrogant or brazen with my 'scientific knowledge'.

Now a doorway was being creaked open and it was clear from Macca's line of questioning that, if I was suitably equipped, an opportunity lay behind it.

Macca's response was clear: "Is there anything in these reports that will tell us how to move forward?"

With a stabbing feeling that I might be decried by the scientific angels who sit on one shoulder, I sided with reality, listening to the devil of pragmatism on the other.

"Er, a bit, but not a massive amount. If you really want to move your athletes on, we'll need to move off this script."

Macca taxed me further: "Then do all the testing, get the reports together, have a look across the twelve guys and give me a summary. Then let's start having a real conversation about what I need to do to start getting results."

You can get along all you like with an athlete. You can get to know them well, you can advise them directly, you can give them your recommendations, but, counter to what many might assume, these are not the primary relationships as an applied scientist. The fact that I had integrated into the whole team was good, but my responsibilities were to the coaches. They are the ones who make the calls on a day-to-day basis, so they are the ones that need to be in the driving seat for the decisions because when the athlete crosses the finish line it's the coach who answers to the result.

Macca wanted more. He had received some criticism for his result at the 1998 Worlds, so he was setting about looking at what more he could do. He was sussing out the options around him, the quality of the information, the intent of the individuals and ultimately whether they could make a difference. Macca's approaches were a call to enter the team, a commitment to contributing to a team goal and giving yourself to the collective. Personal agendas would need to be put to one side and decisions would need to be made in the best interest of the team. Personal ambitions could be served in achieving the team goals, but they were secondary to the overall, primary team goal.

My job was to provide the best scientific support to the British rowing team, so it wasn't an easy decision to make.

I had to step forward. I had to commit. If I didn't, I would be saying, "I don't want to be involved."

However, a transactional relationship feels safer; more stable. As you volunteer to take this step into more immersive involvement, you do so with all of the realisation of how lacking the scientific literature is, how far off-beam sport science curricula are and how much you wished you had read more about what elite athletes actually do.

The rowing physiology body of knowledge was fine at the time, telling us that they are big, have big engines and row a lot. But it hardly mentioned that they improved with training, what the most important training for them to do was or what the most important capability for rowing fast was and what we should measure in line with this.

I had assembled what knowledge I could, and I had received a powerful induction about the contextual understanding of the performance world. Even though I had received this strong warning not to get carried away, this was different. It was an invite. The challenge then was to respond: to be able to apply what I knew; interpret this so it would be useful; and interpret the habits, the methods, the individuals, the responses, the outcomes, and the opportunities so we could make a reasoned judgement and sound decisions that would increase the probability of these guys making progress. I was stepping onto shakier ground.

Macca had his work cut out with the raw material. We all did. The way he handled this situation was pioneering at the time, and is an approach I have rarely seen matched in the UK high-performance system since. He was not taking athletes who should be winning, giving them a training programme, looking after them and then winning. He had a very good bunch of rowers, but under any normal circumstances, with their level of existing capability, and with the normal rates of improvement they should not have won. Macca needed to change the paradigm to exceed expectations.

Macca grew as a leader and it was compelling to follow him as he embraced the expertise around him, galvanised his team and set us on a new course. He engaged the services of: psychologist Dr Chris Shambrook; nutritionist Jacqueline Boorman; specialist in injury biomechanics, Professor Alison McGregor; the late, great Harry Mahon, a tough-talking, but quietly inspiring, coach; and specialist physiotherapists to deal with back pain. We weren't often all in the same room at the same time, but we all knew we were part of a united ambition.

RECOVERY POSITION

Throughout 1999, Macca and I would talk, question, debate and drink tea at various training camps, boat clubs and cafes. He would constantly ask me the same question whenever I saw him: "What's new, Stevie?"

When he first started asking me, it was really engaging. It got my knowledge juices going and made me questions whether I thought an idea was worth proposing or not. It would test my ability to pitch my idea, give him the take-home message, add a technical sprinkle of information and propose a performance benefit.

Macca had a great way of giving me negative feedback. He would listen to my idea before describing it back to me. He would then challenge me to find a way to overcome a certain obstacle, such as how to individualise for a large group of athletes, how to ensure adherence or how to sell the idea.

Sometimes he would say, "No, it just doesn't feel right."

Figure 6. Filling time reading about recovery and adaptation while waiting for
the rowers to row in for physiological measurements.

I had to judge whether, on this upward curve of trust and
growing respect, to push back at Macca; to challenge him
to be more open-minded, for example. Every other week I
would head down to Hammersmith, monitor sessions,
spend time in the launch boat discussing plans and,
despite these knockbacks from Macca, he would keep on
asking, "What's new, Stevie?"

After the initial excitement of having my voice heard
wore off, I started to become a bit concerned because
nothing was getting through. I had opened with my best
ideas, suitably filtered to the top with my own whirlwind
of intrapersonal review, debates with colleagues, reading
and reflection. But I would keep pitching my ideas. After
ten or so attempts of offering what I thought were some
really good proposals, I asked if he was ever going to go
for one of these ideas.

Macca came back in a considered and encouraging way: "Yes, definitely, as soon as you come up with a really good one. You see, I've got to find a way of getting this lot to perform at a much higher level than they're capable of. We need something that will change the game for them, take their performances up and then get them working together so well that the boat really flies. I'm looking to you to play your part in us all being greater than the sum of our parts."

I went home that day knowing that, between the lines, I had been told, 'not good enough yet,' but my overriding feeling was one of inspiration. So I got thinking and decided to go for one of the big principles.

I knew Macca thought the core training programme was sound, but he also knew it didn't quite give him the results he needed from his rowers. Our objective information showed that they could and should have been better. The main area, then, was for us to find the core principles of the programme, but adapt it to suit the athletes vying for a slot in the boat. We also needed to be confident that if we did make changes, there would be a very small chance of any negative effects from the intervention.

Firstly, I had observed on most days that, wherever the rowers were training they would gather, fiddle about for a few minutes, then get into a boat and row for about sixty to ninety minutes. They would then get out of the boat, put it away, get dry if they needed to and have some food, which they called second breakfast[vii]. Within half an hour of second breakfast, they would head to the weight-training room and get lifting before jumping straight onto the rowing machines for a further endurance training

---

[vii] Second breakfasts is a dangerous concept for the support team. It can lead to them adopting the same energy intake without the output, which leads directly to elasticated waistband clothing.

session, which was normally forty-five to seventy minutes in duration. The rowers would be done and dusted by lunchtime, so they would grab a snack and head home.

This sounds like quite an arduous morning, doesn't it? They were tough cookies for going from one thing to the next, you might be thinking. However, this didn't quite make sense to me. It struck me that they were running this schedule more out of convenience, choosing simply to get all the training out of the way early so they could go and do something else with their day, such as studying, working or playing golf.

I raised this with Macca and suggested a change. The idea, which is now commonplace in the world of elite athletes, is that you need to let the training session resonate and take its effect, and for the biological systems to cascade into adaptation. If you want to give the body endurance, give it time to take effect. If you want to give it strength, give it time to take effect; in fact, give strength a bit more time than you do endurance.

The idea was for the rowers to do their usual early morning training and go for their second breakfast (not for support staff; there was nothing more than a cup of tea for us). Now for the change. They would then rest for an hour. This could just be sitting down, lying about, doing their studies or popping back home if it was close by. The hour was there to let the muscle cells, the blood vessel cells, bone cells and immune cells respond, recover and begin to adapt. The rowers would then get back to their training; say, for example, weight training.

The double benefit of having taken the extra rest was that they would be fresher for the training session that followed. They weren't going to be massively tired, but imagine doing a marathon and then doing an hour's worth of weight training. The rowers would be a little way along this continuum. In addition, they would have benefited more from the morning's endurance training.

Figure 7. The GB Men's 8+, from bow (right) to cox (left): Andrew Lindsay, Ben Hunt-Davis, Simon Dennis, Louis Attrill, Luka Grubor, Kieran West, Fred Scarlett, Steve Trapmore and Rowley Douglas.

After this, the rowers would go home and work, rest or play[viii] from mid-morning until mid-afternoon, when they would return to their bases and do their final rowing session, normally on the rowing ergometer. This way, they would benefit from a further four hours' rest, recovery and adaptation before the stress of a further training session would be imposed, from which further recovery would be required, typically overnight.

The two main principles being applied aimed to give the body more time to adapt and, furthermore, to give it a cleaner signal for adaptation. If you go for a run, then

[viii] *This book would now appear to be open to sponsorship from confectionary products.*

immediately do some jumping about, then lift some weights with some stretching thrown in for good measure, the body will get confused as to how to improve. If you were to throw a stone into a still lake, you could stand back and watch the ripple effect wave over the water. If you were to throw another stone in almost immediately afterwards, you would see the ripples from each splash competing and interfering with each other.

The human body responds in a similar way. If you give the body strength training and only strength training, it will have received clear instructions about what it needs to do to adapt: to develop stronger muscle tissue, stronger nervous conduction, strengthened tendons, stronger bone, and maybe a stronger heart and core abdominal muscle. If you give it mixed messages, it will get confused and offer a general, suboptimal response.

The idea took off. Macca could see the logic, which chimed with the scale of ambition he needed to make a difference to the team, and he felt the benefits were worthy of his efforts in convincing them. Crucially, it maintained the same template for training, which meant he was taking on a low-risk, high-return strategy.

Sure enough, it began to pay off. The rowers reported feeling more recovered going into training sessions. They felt less fatigued at the end of a training block and, slowly but surely, their training scores nudged upwards. As their capacities began to manifest, the results began to build. They made a breakthrough that summer by clinching a thrilling silver medal at the 1999 World Championships in St. Catharine's, Canada. The team had developed quickly, but in their minds it was only the first step.

RECEPTIVITY

When I look back at the situation with Macca and his men's 8+ compared to the one with Steve Redgrave, Matt Pinsent, Tim Foster, James Cracknell and their coach Jürgen, what they required of me was markedly different. Redgrave had been at a plateau, for a while, and at thirty-eight he had begun to see a decline in some areas. What Redgrave was seeking was maintenance.

He told me he had tried all types of training. He had seen all the different systems come and go, and therefore his readiness to embrace new, bespoke or different ideas was at a low ebb. That is not to say that the increased recovery idea would not have worked with Steve Redgrave[ix], but at that moment, with all of the information available, with the standard of training and performance he was maintaining, with all of his experience, and with a very high threshold of proof necessary, it would have been difficult to persuade him of the benefits. After all, the chances of him winning were high. The chances of a new intervention working on someone with an extensive training history were poor, so why would he want to change?

Conversely, the men's 8+ had no alternative. Critically, they weren't at the required level, so they had to change. They also had evidence that they weren't responding to training at the rate they might have expected. Vitally for Macca's vision, they were not pushing the boundaries enough to give themselves a chance of reaching the

---

[ix] *Just so you know, the 'increased recovery' did eventually make it into Redgrave's routine. It was put in place for him in early 2000, but was introduced to help manage his sugar levels for his diabetes. Again, this shows the importance of the circumstance and conditions. In this instance, these changed for him, which reset his receptivity to modifying his methods.*

highest possible performance heights, which would go beyond the expected level.

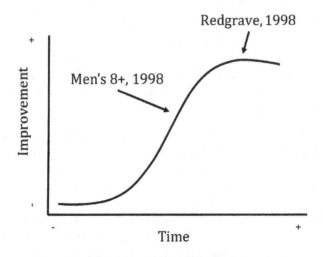

Figure 8. A hypothetical description of the improvement experienced over time with two years to go until the Sydney Olympics. Redgrave was aiming to maintain his abilities, while the men's 8+ had big improvements to make.

On the graph of improvement versus time (above), the 8+ should have been slightly higher. Coupled with the fact that they had all had signed up to an ambitious goal, the conditions, environment and their will made them highly receptive. They had no choice other than to change; the alternative being to continue to stagnate and leave the rest to chance.

The men's 8+ were at the steep part of the curve and the team was formed in this crucible of development. The athletes, coach and support staff were climbing this part of the ascent together. We were acclimatising at the same rate, so the team was pushing each other at a rate that was sensitive to where we all were on the climb.

When parents communicate with their children, their words, enunciation and intonation will evolve as the children grow. No one in their right mind would talk to a baby in full, verbose, erudite sentences. Nor would you talk to a teenager in a patronising, dumbed-down, exaggerated, lip-enunciating way. For a youngster, an explanation will often be necessary for the concept to be understood, while for a teenager a knowing glance could be the most effective form of communication. The interaction between parent and child is hugely contextual to their age and stage of development, along with the maturity of the relationships. So it is with the progress of athlete-coach-support-staff work.

As my relationship with Macca grew, I could integrate additional ideas more successfully. We used breathing training, tempo training and introduced ergogenic aids before the race, to name a few.

Another intervention came in the area of maximum power output. The team had been improving their aerobic physiology nicely. The landmarks of an athlete's physiological profile behave as limiting thresholds, above which things become increasingly hard. Above the lactate threshold, you use more fuel from anaerobic metabolism. VO_2max (the maximal rate at which an athlete can extract oxygen from the atmosphere, transport it in the blood and then use it in the muscle) can only be held so long before fatigue sets in. Typically, this is six to nine minutes. The maximum impulse[x] a rower can generate is a product of the highest force applied to the oar and the stretcher foot plate over the time taken to complete the stroke from the catch at the start to when the handle is by your tummy at the finish.

[x] Often dubbed 'power', but Newton's second law shows us we need to call it impulse (nothing to do with deodorant). Please see reference[34].

I had spotted that the power at VO_2max for the men's 8+ team had been increasing, but the maximum power had not (Tree B, overleaf). Therefore, the percentage of maximum power at which the power at VO_2max occurred was high. On the surface, this seems to be a good thing. But I asked myself, how is the power at VO_2max going to increase any further? Surely the maximum power would act as a ceiling for further development.

The concept got the nod from Macca and before we knew, it in the cold months of February and March 2000, we were constructing resistance training sessions – power cleans, squats and lunges – paired with some 'power strokes' on the ergometer. We were experimenting with heavy lifting, maximal-effort lifting, along with heavy resistance rowing and unloaded, fast rowing.

I think about this example now, and about how much contemporary strength and conditioning practices have evolved to such sophistication of loading methods, creative movement patterns and integrated training that even 'weekend warrior' athletes are now using more advanced methods. What we did in the gym at University College Boat Club at the time would stand up to current methods, but it was certainly 'Heath Robinson' in comparison.

At the time, I was filling in the gap that would later be filled by 'strength and conditioning'. Compared with a team comprising a nutritionist, biomechanists and a psychologist back then, a modern team – blessed with greater resources, funding and an expectation for higher levels of expert services – the multidisciplinary team is far more effective with an array of specialists on board, including biomechanists, psychologists, nutritionists (and chefs), engineers, technicians, performance lifestyle practitioners, ergonomists, doctors, strength and conditioning coaches, skill acquisition experts, performance analysts and data analysts all offering performance-focused support; all overlapping and working together towards the big goals.

Macca and the rowers were starting to gather momentum with the deductive-reasoning approach to seeking out performance gain. We were searching for and finding some of the limiting factors in performance and, from first principles, finding ways in which we could move the squad forward. This concept of limiting and determining factors would later become one of the most important in finding ways of improving performance (more on that in chapter three).

But back in early 2000, this was an edgy concept, certainly when applied to the elite athlete. We were developing ideas from a base of good sense. We had thought the ideas over and over. We were devising novel and innovative ideas that we fine-tuned as we went, turning our ideas into practical sessions and monitoring them to death.

The rowers not only needed to change, but they also needed to have faith in the programme. I had prepped it to the hilt and I was confident that if we could get it right there was performance gain to be had. As it happens, there was. Maximum power went up and, for the first time in ages, power at VO_2max also went up.

Macca had certainly taken a risk. He had trusted me, backed me and given the green light, with the precautionary sense to monitor the responses closely. However, the routine measurements and record-keeping turned the risk into feedback and eventually reinforced the ideas. If I had come up with the idea of maximum power development any earlier, it almost certainly wouldn't have taken off. Any later and it would have been too late. It was hugely relevant to Macca in those few weeks, in that year, with that set of people, and with the way their physiology and psychology responded to a variety of physical and mental stressors.

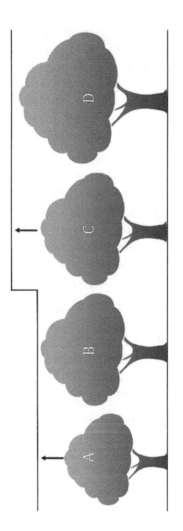

Figure 9. Tree A isn't big, but it has room to grow. Tree B is quite big, but has no room to grow. Tree C is quite big, and has room to grow. Tree D is big and is fully grown. The men's 8+ were B, and we needed to get them to C for them to eventually become D.

There was a clear lesson here about understanding and creating conditions for change before launching them. This is why, when consultants are used in business or in one's personal life to elicit change or to provide expert review, they are more effective if they spend time learning about the environment, experiences and circumstances in the existing team, the industry and the people involved. Their recommendations are then contextually sensitive.

Even more impressive is the use of personal and professional coaching, which aspires to embrace the resourcefulness of an individual or group of individuals in a team. This takes the approach of extracting the necessary information and the best possible solutions from the people with the most information. Often, but not always, it helps to have someone from the outside to facilitate the overall group objective, goal and direction of the team.

Macca set that goal. It was very clear that we all had our own objectives and ambitions, ideas and opinions, but unless we were prepared to lay down our personal agendas, shed our egos and give ourselves to the overall team goal, we would be a negative influence on the achievement of a goal that was worthy of the phrase: 'Greater than the sum of our parts.'

Penrith

Throughout the season of 1999-2000, the team continued to progress in their training, their mental skills, their nutrition, their professionalism and, crucially, their results. They came second at each of the World Cup races through the summer of 2000. When they entered the Olympic regatta, they hiccupped in the first heat. But, as was typical of their attitude and approach, they learned quickly and responded with a dominant display to make the final.

Figure 10. On 24th September 2000, the GB Men's 8+ win a gold medal they had no chance of winning two years earlier.

The pocket power-house, Andrew Lindsay; the gentleman, Ben Hunt-Davis; the smiling giant, Simon Dennis; the uber-talented Louis Attrill; the sardonic Luka Grubor; the confident youngster, Kieran West; the unrelentingly sarcastic Fred Scarlett; the calmly determined Steve Trapmore; and, last, but by no means least, the crazily charismatic Rowley Douglas, took their chance on 24th September 2000. From the very first stroke of the final and for the next five-and-a-half minutes, they delivered as a team, winning the Olympic gold medal on the lakes of Penrith in Sydney.

I, on the other hand, stood in the stands with tears running down my cheeks, humbled to the core by their collective achievement.

Working with the men's 8+ under Macca's leadership taught me:

- To let go of the things I had been clinging to that were peripheral to the main effort.

- To sign up to the big goal and channel my resources towards it.

- To embrace the team and work *with* people. Individual contributions can be effective, but they cannot match the powerful collective effect of teamworking.

Figure 11. Olympic champion coach, Martin 'Macca' McElroy, with wife, Grainne and daughter, Orla, immediately after the win. When interviewed by the BBC, an emotional Macca simply said, "I'm so proud. So proud."

Chapter 3: How do you Know?

"The first point of wisdom is to discern that which is false;
the second, to know that which is true."

Lactantius

Determinants of performance

I have a confession to make. I cheated in my first year A-level biology exam! Terrible, aren't I? But it's all right. It's all sorted. I have confessed and been forgiven by the teacher.

Being a diligent student, I had popped along to see my inspirational biology teacher, mentor, training partner, colleague, co-author[4] and great friend, Colin Clegg, to ask for a recap on a particular topic. I knocked on the staff room door and poked my head around it. I saw that Colin wasn't there, and that none of the other teachers were either, so I thought I would just leave him a note.

As I was rummaging in my bag for a notepad, I glanced across at his desk and saw a pile of freshly photocopied exam papers. I tilted my head to peek at the top page to find out which unfortunate group this stack lay in wait for. My eyes bulged with surprise when I saw 'Biology A1'!

After a moment's pause, I thought, 'No, I can't.'

I slowly proceeded to get my notepad out. I tore a page out along the coiled spine, wrote the note I had planned to write, folded the book closed and popped it back in my bag. Then I looked around. No one was there.

I thought again, 'No, I can't. That would be wrong.'

Then, in a flash, I thought, 'I bloody well can,' and in the fastest movement I have ever achieved, I whipped one of those naughty, temptress exam scripts into my bag. I immediately felt bad, but not bad enough to resist devouring the paper over the next two weeks.

Now, the script cover page had no date on it, so I couldn't be 100% sure it was the right one. Therefore, I devoted some time – maybe 10% of my revision time – to studying other areas, but I drafted and practised every question in the paper to my fullest capability. I remember many of them well to this day, so drilled were they into my memory! Exam day came, the papers were handed out and, as I caught sight of the familiar front page, I let out a little embarrassed whimper. I romped through the paper like a man possessed.

Guiltily skipping home, it suddenly occurred to me that I might actually get 100%. Then a sense of doom rushed over me... 100%! That wouldn't be good. In fact, 90 or 95% would be ridiculous and 85% would be suspicious. Then it dawned on me that I could get rumbled!

Results day came and I had already prepared my admission of guilt to Colin. I quickly scanned the noticeboard for my result. I didn't have to look far; I was ranked top of the class. My score? An inconspicuous 69%! My first reaction was relief that I wouldn't be interrogated, while my second reaction – all played out in front of my classmates) – was disappointment. Only 69%? I had normally batted around the 55% level, roughly in C-grade territory. Now I was a B+, just shy of an A, with a fortnight of cheating under my belt, having seen the actual questions!

"Congratulations, Student Ingham," Colin praised. "Quite an improvement there. I am really proud of your efforts."

'Oh no!' I thought! 'I've just upped everyone's expectations.'

Colin handed out the marking scheme and I headed home to pour over its contents, just as I had with the exam. I found a little error here and a missed point there. I had lost points on my short essays towards the end of the exam, struggling with the compulsory forearm fatigue from writing at the speed of a seismograph needle recording an earthquake!

I began to realise there was more to exam performance than seeing the questions in advance. There was a performance aspect to recalling all the necessary information, but there was also a performance aspect to getting it all down and packaged on the big day. Essay writing, forearm fitness, spotting exam question keywords, matching answer content to the points available, and of course knowledge recall – and especially knowledge recall under pressure – were performance determinants.

I set about not only studying for my exam subjects, but also rehearsing examinations. I got a B in the end; not bad, given the academic negligence and underachievement I had shown throughout my time at Oakmead Secondary School. From then on, I was driven by a desire not to disappoint Colin, the man who had so inspired me to learn, and driven by the realisation that generic preparation for any task would not do. You must work at the components that give you the most success.

STEEPLECHASER

As an athletics fan, I already knew of Mark Rowland before I met him. I had a fond awareness of his style, his character and his results. He was a gutsy, determined runner, and he would happily let his jaw do whatever it needed to do while he gave his all to a race. His post-race interviews were full of exhaustions, passion, fast-talking and, unlike today's media-trained types, he shot from the hip.

He was also a steeplechaser, which puts him in the same category as 400m runners, 1,000m ('kilo') riders and 200m butterfly swimmers. Among sportspeople, these events have a special section of their own, simply labelled 'hardcore'. Rowland was also successful, winning an Olympic bronze medal at the 1988 Seoul Olympics.

His post-Olympic final interview should tell you a lot about Mark's honest and direct character: "I just kept telling myself to dig, dig, dig—and I did it! I've got a bloody medal!"

He came to see me in the spring of 2002 looking for some support for his two main athletes: Hayley Tullett and Mike East. Hayley was a regular fixture on the international scene, having made the Olympic team in 2000, and numerous indoor and outdoor performances. Mike had just broken through. In a thrilling sprint finish, he had nabbed a bronze medal at the European Indoor Championships in Vienna.

They had originally received some physiological support from another laboratory and had grown frustrated with the slow turnaround of results. I had always set up my analysis systems to plot and print out an athlete's data sheet and report in the time it took the athlete to cool down. This was specifically designed to capture them at their most receptive and most interested, while they were still captive in the room.

Mike, Hayley and Mark had been forced to wait six months for one scientist to provide them with their data feedback! Even then, it was a myriad of messy tables, incomplete data and, all-in-all, it was an embarrassment to the profession. The most incredulous aspect of the data presented by our tardy physiologist was the inclusion of a whole host of irrelevant data.

Effectively, as Macca had previously homed in on with his "So what?" question, the data you choose to present to an athlete needs to bear some sort of relevance to their mission. If you report body fat and it goes up to an excessive level, you should (although it's uncomfortable to do so) have the conversation about not lugging that extra dead weight around. In Mike, Hayley and Mark, we had a set of disillusioned people who weren't sure whether they should be trying to increase their respiratory exchange ratio or decrease their maximum heart rate.

Within a few minutes of them stepping off the treadmill (bonus points for me), we sat down with their most relevant data printed off. I proceeded to give a commentary as to what was up and what was down, and what each measure meant.

This was greeted with, "It's good to hear what all of these things are." (extra bonus points for me.) With Macca's question – "So what?" – at the back of my mind, I ventured a little further.

"You're probably wondering, 'So what?' Well, out of all these areas, your priority for development is... your lactate threshold."

Figure 12. Mark Rowland coaching.

Mark looked over at Hayley, who returned a wondering glance, which read, 'Are you thinking what I'm thinking?' followed by a 'What do you think?' from Hayley.

A pause, an adjustment of the coat, a scratch of the chin and a shuffle of the report papers later, Mark said: "Well, it's just that we've been told that before by a couple of physiologists, but we don't buy it. Hayley's threshold, her tempo running, has always been her strength."

I responded as best I could: "Err, well, it hasn't gone up much and it's at a low percentage of her VO$_2$max."

"Yeah, that's what the last ones said, but we still don't buy it."

Then came the killer move from Mark. "How do you know that threshold is the thing we need to work on?"

I was beginning to sense that I was about to be shifted into a new way of thinking again.

"Err, well, you see, your lactate threshold is normally at 80-85% of VO_2max and Hayley's is..."

"Yes, yes, I hear that, and I expect having a higher percentage is a good thing, but what's that based on? Is it based on a bunch of college students, some average runners who are never going to be any good? Or on a group of runners from a different country who don't train like us?"

It was a devastating line of enquiry.

As the questions ran through my head, I could think of the concepts and papers that promoted them, and the original papers on which they were based. As I did so, the foundation upon which I was making my recommendation started to rumble and shake.

I was caught in a circular argument here: 1) Yes, a threshold at a high percentage of VO_2max is *probably* an advantage. 2) This hadn't really been shown to be a normal observation in highly trained middle-distance runners, especially those with a very high VO_2max like Hayley's!

I had to concede, "It's probably an advantage, but you're right. The basis on which I'm suggesting that you work on this is not particularly specific for you, your training background and your physiology."

Whereas Macca had driven hard with, "So what?" Mark had uncovered me with, "How do you know?"

It had exposed me, my knowledge and many of the tenets my profession proposed. It stripped back the certainty with which I was making my prescriptions and, unlike with Macca where, in response to "What's new Stevie?" I was suggesting ideas based on individual papers or a single line of investigation.

Here it was: the Chinese whispers following a line of publication to citation to interpretation to recommendation to prescription to promotion of an approach. At any given point in that chain, one step wouldn't be a wild leap away from the next, but from one end to the other, the validity and applicability of interpretation would be a far stretch. However, in my current position I realised I was liable for promoting an idea that, at best, could be right, but the foundation upon which it was built was *probably* weak. At worst, it had become distorted and magnified over time.

THE MODEL

With Mark and Hayley (not so much Mike, as he was much more chilled out), the prospect was clear: they would need to have far more confidence in my knowledge before they would accept my recommendations.

Ruffled by another intuitive line of questioning, I was left with several fundamental questions: What is the actual basis of my recommendation? Is the evidence base strong enough for me to act? Have others drawn success from advising similar suggestions, albeit from a questionable evidence base? How could the evidence base be better? How are Mike and Hayley actually hitting these areas in training?

I decided to adopt a two-pronged attack. First, I delved into some data to see if I could improve my knowledge base. Second, with Redgrave's words ringing in my ears, I took the approach of finding out what they were actually doing and why. I suggested to Mark that I should spend some time with them to see how they trained.

Delving started with a deep dive into the archives at the BOMC. This might make it sound like I was in a panelled library adorned with Olympic memorabilia and signed sportswear from former champions, and with Vangelis' *Chariots of Fire* instrumental theme playing on a continuous, rousing loop.

The reality involved going through old school lockers in a dark cupboard at Northwick Park Hospital. Northwick Park was certainly a strange place to find an Olympic Centre. The hospital would rumble along with its daily proceedings and then the occasional legend of British sport would wander in to visit the dusty laboratories and clinics, seeking support or treatment from the tens of staff hidden within.

The BOMC was a world away from the current British system with its integrated force plates, alter-G treadmills, immersion water therapies and, more importantly, the integrated, multidisciplinary, individualised, problem-solving support teams.[xi]

I had visited the centre in 1992, six years before I ended up working there. Colin Clegg had arranged a sixth form field trip visit to the centre and I had gone there with a mission to find out whether its staff had spotted a disproportionate increase in the abilities of some athletes in order to see whether there was any evidence of doping![xii] I was interested in this area for a project and it was a reasonable question. However, you should have seen the look on the face of the physiologist who showed

---

[xi] *I can clearly recall taking a shot-putter through a VO_2max test because that was what was expected, with our bog-standard protocols. I knew then and I cringe about it now, as it clearly wasn't of any real use compared to strength assessments.*
[xii] *Twas very fruity of me to be so bold.*

us around that day with this hot-headed eighteen-year-old asking about the probabilities of detection and the ethics of doping.

In exchange for the tour, we were participants in their reliability trials and I remember sitting in the isokinetic dynamometer chair ready to have my hamstrings and quadriceps ratio quantified. I gazed out of the window and saw a game of football taking place in the fields just opposite. I wondered which Olympians these might be! If my doping questions had the penetrative enquiry of a journalist, my daydreaming showed the naivety of an imp.

Legends of sport science and medicine, Professor Craig Sharp and Dr Mark Harries, had established the centre in 1987 in response to criticism regarding the poor medical care in the lead-up to the Los Angeles games in 1984. Upon receiving the keys to the old hospital rooms, Craig and Mark had reportedly opened the doors to find several hundred Zimmer frames and walking sticks in a pile.

In all his grand witty wisdom, Craig commented, "Has Jesus just been here to heal the lame?"

But I digress. I continued to rummage and rummage through the files. I was intent on finding athletes who were similar to Mike and Hayley; ones who could hold a candle to them; ones who would receive at least a shrug and a nod of approval from them. Steadily, I began to accumulate a pile of decent physiological specimens labelled under 800m/1,500m and 1,500m/3,000m.

Then came the sorting, first – as a jolly good scientist would – to look through the background data sheets to see which analysers had been used for each of the tests. It's easy to assume that all of these machines we huff and puff into or insert our bodily fluids into are pretty much the same. Yet, bafflingly, they are all pretty much different and carry out the analysis in different ways.

The second phase was to work out which were credible athletes, which meant scouring through the various, and at the time paper-based, databases to ascertain whether the athletes had registered a decent performance time. Although it is certainly not ideal in a pack running sport, where the races can be tactical (i.e. slow, then fast; or fast, then slow due to fatigue), jostling for position and involving the need to run a little further than necessary to stay in position, the personal or season's best times represented the best available data.

Individual time trial performances would have been much more reliable and robust, preferably on a 400m indoor track, but such events with any of these components just don't occur. I plumped for 15% of the British record, which was none too shabby at the time, given the long and rich history of Sebastian Coe, Steve Cram and Kelly Holmes clocking world records or world-leading times.

I then filtered a bit further, looking at when the physiological profiling had taken place and whether there was a match to a running performance. It's no use taking a cross-sectional record of an athlete's physiology from, say, November 2012 and then matching it to a running performance in May 2016. They could, and often would, be different physiological specimens and different athletes at these points. They would also be different because of the type of training they would be doing at different times of the year as they prioritised endurance early in the season and race-specific training during the competition season.

This crucial last step was important to ensure that the matching of physiology to performance was legitimate and credible. I set a four-week window either side of the performance date to allow me to match my data. The figures were then sieved through these various layers of acceptability, all designed to ensure that the question of 'How do you know?' could be defended from a firmer evidence base.

Eventually after about four months of beavering, away, a table of information – confidentially coded by athlete, date, performance time and a host of physiological test data – was collated. I now had a fabulous set of data I could share with Mark. I could give him a robust comparison point, whereby he could see where his athletes sat in relation to other British athletes. This was even more compelling for Mark because, although I couldn't share specific personal details with him, he knew that this group of athletes included many against whom Mike and Hayley would compete. The discussion reinforced my view that the lactate threshold was indeed in need of some work as they didn't compare brilliantly with their peers in this area.

The next step was to gather some stats. Here I brought in Professor Alan Nevill, of the University of Wolverhampton; one of the world's leading biostatisticians. He undertook some mathematical wizardry to explore which of the physiological factors were discriminatory. This step was crucial in our understanding. Prior to this, we had only a shaky evidence base. We had some data that seemed to support our idea, but ultimately it didn't hold up to analysis that essentially asked: On any given day, can you do without that index? Are other factors more deterministic?

The results showed that the threshold positively correlated with performance. The higher it is, the greater the running speed. This made sense, and again supported the proposition. However, under the hot spotlight of unforgiving mathematics, it didn't get selected. It was spat out. Instead, the model chose a composite of VO_2max (which makes sense) and running economy (we haven't met this one yet, but it's the ability to run quickly, while using less energy to do so, which makes sense).[5]

The statistical analysis effectively said: "Hey lactate threshold! You're nice, and I can see that you're a good dancer, but everything you contribute to the party I can get from those two amazing dancers over here: VO_2max and economy. They're the life and soul."

It simply wasn't one of the critical factors. There are always errors in such models with natural variation, inter-individual differences and unexplained abilities (the big one here being anaerobic contribution to performance, which has long plagued applied scientists as it is very difficult to get a handle on), but it was a huge step forward in terms of clarity.

The method would nowadays be labelled data analytics; an area that celebrates the grand elegance of concluding from authoritative data sets, but one that has become flooded with hype and, most frustratingly, jargon. The determinants of performance, have become known by several terms: constituent factor analysis; technical performance indicators; correlates; critical success elements; component factors; constructs; discriminators; contributor analysis. They all describe the same thing: the factors that determine performance.

The approach we undertook and the model that followed was one of the strongest steps I had ever taken in offering confident advice to athletes. It did, however, mean taking a big step backward before properly moving forward. The brave move Mark took had shown me it was necessary to reject previously passed down dogma to develop our own story, our own evidence base; one whose origins we knew and in which we therefore had confidence.

It would have been all too easy to jump the gun, for Mark to have bowed to the science, but with just a few questions he had unsettled me into thinking more critically. This gave us both a framework for our thinking, decision-making and subsequent actions.

In modern-day applied science, there is a phrase that affirms the confidence to use science: "Why guess, when you can know?"

I am told by Baroness Sue Campbell that she had borrowed this question from Michael Schumacher's race engineer and used it as a mantra to establish the need for objective services in elite sport. The phrase stretches us beyond the need for a randomised control trial at every corner. It compels the recipient of science to understand what is going on, which gives them the confidence to move forward, adapt and ultimately progress.

'Knowing' also calls for the patience and diligence to invest the time, effort and invasiveness of taking additional measures, having samples taken, listening to the interpretation and discussing this in new depth. Mark had reflected his coach's version of thinking in this way onto me. The simple question of "How do you know?" had pricked the arrogance bubble of applied science and given me a wake-up call.

Mark had guided me to a stronger philosophy. He wasn't willing to just copy what had been previously advised. He was challenging me to know for myself. His questioning led to uncertainty, which led to our running model, which led to a reference point, which led to decision-making in training and performance, all of which were individualised to Mark's athletes. Beyond establishing an evidence base for the physiology of middle-distance running, we had started to cement a way of working that was curious and critical; that questioned and explored everything. It was truly the first time I had worked with a coach and athletes who were willing to take a 'no stone unturned' approach.

COACHING THE COACH

Figure 13. Mike East in full flow.

You're probably wondering where this hotbed of analytical, philosophical and scientific revolution was going on? Monaco perhaps? Lausanne maybe? Somewhere exotic on a training camp, you might expect. Well, not for us! The Spectrum Leisure Centre in Guildford was the microcosm for our curious, critical and innovative thinking.

I would head down there every other Monday with my colleague Kate Miller. We would pack our analysers, latex gloves, capillary tubes and so on, taking care not to forget headache tablets!

We were working on fine-tuning Mike and Hayley's tempo running, which typically comprised a twenty-five-minute session of 1,600m reps. Most athletes will overcook this type of session, choosing to see it as a series of 1,600m all-out efforts. This is all very well and it may help some people get fitter, but Mark was aiming for this session to challenge the athletes without overloading them; to give them a dose of difficulty, but one they could cope with. The aim was to ensure that lactate production was just about matched by lactate clearance, which gave a nice stimulus for the muscle cells to improve. A bit like Goldilocks and her porridge sampling, any easier and it would have been too cold (for this area of fitness to improve); any harder and it would have been too hot. If it wasn't too hot or cold, it was 'juuuust right'.

Eventually, after about a year, we convinced the pair of them to stop running off in search of hot porridge, and to calm their pace to the levels Mark was hoping for[6]. Pleasingly, adhering to Mark's programme as he intended it, led to good improvements for both athletes. Given what we had learned about the importance of lactate threshold, this wasn't me finding a way of addressing it through the back door. As any sensible coach would do, Mark had a certain type of session designed to address the lactate threshold.

What we found was that Mike and Hayley were running it in a way that didn't match with Mark's requirements. In fact, and in keeping with our running model, once we had optimised the 'tempo' sessions, they both ended up doing less of this type of work, which was exactly the type of priority it deserved and required, given that other abilities were more important. This led to an alleviation of unnecessary stress in low- to middle-intensity training and allowed more emphasis on the high-intensity sessions, giving Hayley and Mike more gas for the high-intensity training sessions, which were more specific to their racing physiology.

After finishing the tempo session, we would pack up and go for a coffee at the leisure centre. Initially, the coffee meeting would be there to debrief with the results, but as the relationship developed, Mark and Hayley would fire question after question at me and Kate. Mark was incredibly open and would bare his soul about his thoughts in a stream of consciousness (for full effect, read the following as quickly as possible).

"I was thinking about 400m reps, but then I've been reading about a nice session with 500m reps, although I really like the idea of doing some pyramids going from 300m to 600m, then back down again to get some variety in there, but it's less of a known quantity, but then I'm not sure what recoveries to use. Should I send them off on 60 seconds for the 400m reps, or maybe 90 seconds for the 500m reps? If I use the pyramids, should I keep the recoveries the same or adjust them to go up with the rep distance, or should they go down so they can really get into the shorter reps?" (Looks up at me.) "What do you think?"

This would all be expelled in one breath.

This is why we needed the headache tablets. I would be processing so many pieces of information at the same time, trying to evaluate them all as they came flying out. I would be weighing each of the options up and then trying to give an answer. It frazzled my brain! At first, I attempted to address each of the questions one by one, but as the answers came out they sounded fudged, mainly because they were.

So, after a while, I learned to let Mark offload his machinegun on me, and let the headache threaten, but then I would turn the situation around so I was the coach in this instance. I would ask him, "What is it you want to get out of the session?", "What do you think the guys need at the moment?", "What do you think is the strongest option?", "What are your options?"

Often by this point in the conversation, Mark would say something like, "Yeah, yeah, I'll go with my first option. That's really helpful, thanks." So, in reality, Mark had the answers himself. Giving him an outsider's view would not have helped greatly, because there was too much information and Mark had already been given it all. He needed to make the decision himself because he held the most information overall. I would call this 'coaching the coach'.

ICED ATHLETES

Hayley was one of the most critically minded athletes I have ever met or worked with. Everything she did, and anything on the radar, attracted the full force of her attention.

A nice example of this critical thinking centred around the use of ice baths. This aquatically masochistic craze emerged into popular practice among athletes in the early 2000s, all in the name of recovery. It was taken up most vociferously in track and field athletics through one of the leading proponents, Paula Radcliffe. Jumping into a bath, wheelie bin or specialised inflatable dinghy-type thingy is still common practice for many, and it is something I continue to recommend for top athletes.

However, I only recommend it for in-competition or post-competition recovery, where a quick turnaround of recovery is the priority. You need to immerse the limbs and body parts that need recovery in water at a temperature of approximately 10-15°C. You don't need the water to be ice cold, although knowing that doesn't stop many taking the 'more is better' approach and trying to get into a block of ice!

However, encouraged by the supporting cast of physiologists, and strength and conditioning coaches, many athletes took to the practice on a near daily basis, particularly after 'hard' sessions. These hard sessions were typically the Tuesday or Saturday/Sunday interval/repetitions sessions. In the British systems – built logically around a competition schedule – Tuesdays and Saturdays are normally very hard. If an athlete is going to puke at a track session, it would be on these days. What you can guarantee, particularly early in the season, is that the following day they'll have some stiffness or soreness, originating principally from the impacts and associated eccentric (stretching) muscle actions, the extreme lactic-acidosis experienced in the muscle, some swelling from inflammation and some glycogen retention. Wednesday sessions would normally see many people hobbling around and generally going slower. Ice baths seem to temper this response, and while the athlete will probably still be sore, the pain is often reduced.

Hayley started to use ice baths in 2001, and had been doing so for about six months when I met her. After a full season had elapsed, she sat me down and said: "Look, I'm doing my ice baths, I'm feeling fresher the day after a hard day's training. In fact, I'm feeling fresh enough to do some additional miles. For the last few years I've hit about 70 miles (113 km) each week, but now I'm doing 10% more for the first time. All the signals are generally positive, but I don't seem to be any better off. If I look through my training diary at all my track sessions, my times, my recoveries" – [and believe me the diaries were as thorough as I have ever seen] – "I'm not any faster at any given point. Am I just doing more to get the same effect?"

Once again, this line of logic had great sense, clarity of thought and critical thinking infused through it. The body is smart. If you give it a stimulus to change, it needs time to let that stimulus resonate, as we talked about with the men's 8+, but if you present another stimulus it can get confused. With the presentation of cold through ice baths, you are interrupting and blunting the cascade of response and so there is every chance that the mechanisms of adaptation will simply shrug their shoulders and not bother to improve nearly as much.

For an Olympic athlete, there are always going to be goals to accomplish each week, month and year, but, ultimately, it's all about qualifying for and performing at the Olympic Games every four years. So it is the long-term adaptation that should be prioritised, not whether you feel fresh the next day.

Ice baths and other recovery methods are all still hotly debated. However, in a sport often blighted by impact-based soft-tissue injuries, increasing the amount of training you are doing without any further return is not a sensible move. If anything, you really want to do less for maximum gain. It was this type of penetrating intuition and dogma-busting questioning that most brilliantly typified Mark and Hayley's approach. It was certainly the strongest case I experienced of a coach and athlete who were prepared to question everything, even if it meant me stocking up on paracetamol to keep up with them!

Figure 14. Hayley Tullett's last competition at the Melbourne Commonwealth Games in 2006, where she took bronze (reproduced with permission from Welsh Athletics, copyrighted).

Hayley and Mike would go on to achieve very high levels of performance: a sub four-minute 1,500m for Hayley and a 1,500m at three minute and thirty-two seconds for Mike. Looking at the doping scandals that have since come out in the press and the courts, I am sure there are many who would feel Hayley's incredible bronze medal at the 2003 Athletics Worlds Championships should have been a higher rank; or that perhaps she should have been declared the outright winner. Mike came sixth at the Olympics in 2004, which was an excellent result against some of the best middle-distance runners of all time. Hayley and Mike certainly peaked in those respective years and importantly progressed at a rate that broke the normal plateau in performance seen with maturation.

Looking back on my time working with Mark, Hayley and Mike, I learned the importance of:

- Curiosity and allowing critical thinking to predominate over accepted dogma.

- Developing an evidence base for the questions in hand to provide a concrete reference point from which we could fine-tune our focus.

- Listening to and respecting athletes' views and opinions. Without their questions and thoughts, I wouldn't have been able to give them the best possible support. In turn, without their questioning, I don't think they would have got the best from me.

CHAPTER 4: SEVEN SPINNING PLATES

"Reading, after a certain age, diverts the mind too much from its creative pursuits. Any man who reads too much and uses his own brain too little falls into lazy habits of thinking."

Albert Einstein

ICED RUNS

Toni Minichiello is the coach with whom I have had the longest working relationship. It stretches back to 2001 and is still ongoing. Our paths first crossed when his wife, Nicola, a talented heptathlete at the time, trialled and made it onto the bobsleigh team. She made an immediate impact, even though she hadn't been down a bob run before. I met her in the labs and put her through her paces and, on our all-out sprint cycling test, she came out top, as she did for the sprint tests, the lifting tests and so on. Suffice to say, she was very capable physically.

I travelled with the women's team to Utah, USA, in early November 2001. A month earlier, I had accompanied the men's team to Calgary, Canada, with the same objective: to provide them with some answers about how their physiology responded to altitude and what they might be able to do about it. The team were great fun to travel with.

It was the first time the women were to compete at the upcoming Salt Lake City Winter Olympics in 2002, and the bobsleigh track was in Park City. It was a great camp with some excellent improvements for the girls, but with one notable experience to boot. The daily routine with the team, was a leisurely start – compared with the eye-watering 5am starts with the rowers – for breakfast, before heading to the bobsleigh course, taking the two runs permissible for each team, heading back for lunch, resting, then either doing weights or sprint training, and ending with a nice relaxing evening of sanding and polishing sled runners!

During the camp, there was talk of the teams' sports massage therapist taking a slot in the back of the bob with the head coach of the US team, who had designed the course. I was keen to see him go down. The ten or so girls on the team gathered around to watch him descend and the US coach came over and said, "Yep, we need to test the timing system. Are you up for the run?" before thrusting a spare helmet in his general direction.

His head retracted. He was visibly startled and, with his head shaking from side to side, he blurted, "No... I don't want to."

'Oh, that's a shame,' I thought to myself. 'I thought he'd been looking forward to it.'

Just as my eyebrows lowered, I saw the crash helmet swing towards me and stop by my chest. "Do you fancy it, man?"

A huge upsurge of adrenaline immediately took hold. Firstly, I was surprised. Why hadn't I thought through the fact that the intended novice bobsleigher might not fancy doing it? Then in the space of a microsecond I had a prolonged conversation with the angel and devil on my shoulders.

It simply went, "You gotta do it. That's a once in a lifetime experience!" versus "Don't do it! It'll be hideous," repeated in cycles about a hundred times.

The naughty devil won that morning, and before anyone could say, "Feel the rhythm[xiii]," I had blurted, "Yeah, alright."

I had the helmet fitted and was quickly concussed with the numerous head smacks of approval from the team.

I didn't do the run and jump-in sprint start; I simply sat at the back while the US coach said, "You can watch the first five turns."

After the first rough, bumpy turn, I decided head down was a better option. By the end of the run, I had a newfound admiration for this category of sportsperson who can launch and drive a bob at 130 km/hr, withstanding the feeling of a small elephant sitting on their back while bodily fluid is pressurised out of their noses, with ringing in their ears, and being shaken to within an inch of their lives. That's bobsleigh in a nutshell. Both angel and devil had been correct. It was a once in a lifetime experience, but it was also hideous.

[xiii] *See the film 'Cool Runnings' for an explanation of this cultural reference.*

THE GODFATHER

Working with Toni back in 2001-2002 to support Nicola and her transition from track and field athletics to bobsleigh, I could tell that he was a hungry, intuitive and bold coach by the line of questioning he would take with me.

"Why circuit training? What's the point? It's so unspecific! Is it just training the mind to be tough? Can you get the same effect by doing more reps in weight training or medicine ball work?"

Toni has become known as a very outspoken coach, and on any given day it's possible for him to tell you he can run British sport in three days a week and spend the rest of it on the golf course. He doesn't lack for opinion!

We worked through several scenarios for Nicola, which helped her get into a good enough condition to progress onto the team. She eventually became World champion in 2009.

Within just a few months, our paths crossed again via an athlete named Louise Bloor in 2003. I was the physiologist on the BBC show *Born to Win*, searching for the next champion, but unfortunately it required the winner to be good at everything, rather than to be amazing at a specialist event, as is required to excel in sport. Still, it was a nice three weeks in St. Anton, Austria!

Louise won the show and it was my job to provide advice and training to all the contestants. Louise is a fantastic athlete and went on to become a multiple AAAs champion and 4x100m relay team member for Team GB. She had been well developed for *Born to Win* as a result of Toni's help.

As I continued to support Louise, Toni and I chatted a lot about methods, underpinning knowledge, and all the whys and wherefores of training. He also started to ask me questions about another athlete, one I had already heard of in athletics circles and one whom Toni clearly knew would be special.

"What do you know about heptathlon, Steve?" he asked.

"I could tell you the events in order, but if you're asking for the science behind it, not a great deal," I admitted.

"Alright, what about 800m running?" he probed.

"Well, I can help you with one of those," I replied.

In early 2005, we sat down in earnest and Toni spelled out his position in a mouth-wateringly appetising, ambitious and open-minded way,

"You know I coach this kid, Jess Ennis. She's good. She's *really* good. I think she'll win stuff. In fact, I know she will. But I want to get it right and I need some people who can help me. Would you be up for it?"

"Of course. I'm happy to help, and it's my job to do so," I told him.

"Well, if you're up for it, I think we need to do something new and innovative. Something no one has ever done before."

I had encountered some substantial challenges from coaches and athletes thus far in my career: "Are you going to make me go faster?" "So what?" and "How do you know?" These challenges had reset my thinking, galvanised my efforts and driven me to find out for myself, but this challenge was different. Let me just take you through those words again...

"Jess Ennis. She's good. She's *really* good. I think she'll win stuff. In fact, I know she will. But I want to get it right…"

This statement showed a real maturity of outlook about the prospect of handling this rare talent. When Toni had first encountered Jess, she had made a huge impact by beating not only everyone in her age group, but much older girls and boys. Toni had coached Jess to break junior record after junior record: local, regional and national.

With a talent like this, it is incredibly difficult not to live in the now. The immediacy of wanting results and titles right away is a natural impatient urge. Wins enable you to gather influence and have your say on what should be done. He had the sense to know results would come as her natural ability was strong enough to give her success. Toni's vision spoke of 'getting it right', which was a vision of looking back from a future standpoint and seeing that he had done everything he could to optimise Jess' career. It was an uncommon sense of patience, care and honour that Toni held dear in his duty as Jess' sporting guardian.

Now on to the next half of his statement: "…we need to do something new and innovative. Something no one has ever done before."

What a statement. What a challenge. What an inspiring, motivating and embracing 'call to arms'! He was accepting the sheer complexities of the heptathlon; the quagmire of events. He wasn't viewing this as insurmountable, but he wanted to break the mould, cast a new dye and get results beyond the athlete's capability. He wasn't necessarily looking for a revolution of 'fire', 'the wheel' or 'the 'Fosbury Flop' proportions; he was simply talking about a new approach. He wanted to deconstruct the heptathlon, to know everything about Jess and then to undertake a perpetual examination of how we could do things better.

This was "So what?" multiplied by "How do you know?" multiplied by, not only, "Are you going to make me go faster?" but "Can we go faster at the right moment, perhaps eight years in the future?" (Not as catchy, but certainly compelling!)

I had never before encountered a conversation quite like the one I had with Toni in early 2005. He had the athlete, the perspective, ferocious ambition and deep care for Jess' prospects. He had spent eighteen months sounding me out, quizzing what I knew, querying how I would make it come to life for his athletes and he was finally ready to take me on.

If Toni has an outstanding characteristic, it's that he values loyalty to a Corleone family level. He is fiercely defensive, protective and, most particularly, inclusive and exclusive in his approach. As our relationship developed over time, he would ask me questions at a depth that showed his willingness to be completely honest and, at times, vulnerable. It was highly consultative, and he would reveal his gremlins, his ideas, his concerns and his hopes, bearing in mind that this is a man who is also brash, confident, energetic and at times quite noisy!

On board, and with a sense of innovation, ambition, openness and keenness to dance with all the complexities of heptathlon, I opened with a big question to start us off: "When do you want Jess to peak?"

Toni didn't hedge his bets too much: "The Olympics in 2012. Hopefully that will be in London, but I doubt that very much!"[xiv]

I pursued this line of thinking further; in part to gauge the extent of his thinking, and in part as a way of sticking

---

[xiv] *How wonderfully wrong he was. It would be eighteen months before he would realise it was indeed to be in London, and eight-and-a-half years until he would realise his own plan.*

a toe into the water to see whether his open, innovative campaign was legitimate.

"Okay, so what are you prepared to concede between now and then to maximise your chances of succeeding in 2012?"[xv]

This question tested his faith, patience and clarity of purpose if Jess was to peak at a certain time. Up against this ambition is, of course, the very real possibility of not making the Games and therefore not being able to peak at all. We would see this play out for Jess in a very real sense in her career, particularly in 2008 ahead of the Beijing Games. The question is, in a sense, a test of ambition to summit, but also an awareness that the day you want to peak might not be possible or available. It's bravery versus safety; spreading your bets versus going all in.

Toni looked for clarification as to what I meant. I explained that I was questioning what he was prepared to compromise on in the short-term to gain the most in the long term. Heptathlon is a case of spinning seven plates continuously. The athletes will have several plates they really like and spin beautifully, and others they don't like very much, but with which they need to fall in love or the whole performance will be horrible. If one plate was allowed to wobble, fall or smash, the whole performance would be a disaster.

Sunday roast

In physiological terms, heptathlon is a menagerie of challenges. It requires all physiological systems to be in tip-top shape, to be able to run fast with things in the way,

[xv] With no sense of understatement, I believe this type of question is the epitome of the morally robust use of applied science to aid performance, and the inverse of short-termism often seen in the morally absent world of doping.

jump up high, throw a heavy thing, run fast again, jump far, throw a pointy thing and finally run quite fast for an uncomfortably long distance. Never mind plate spinning, this will get your head spinning!

If you're working with a marathoner, you give them endurance work. If you're dealing with a weightlifter, you work on strength and high force/dynamism. Genetic predisposition means those operating as elite athletes in these physical sports have extremes of physiological type. This could come in the form of the frame, lung and heart size, e.g. Redgrave and Pinsent, but relative to their body size (capacity divided by body mass), and, just as impressively, Mo Farah, Chris Froome, Laura Trott and Paula Radcliffe; but it also needs to take into account the form of muscle morphology, i.e. the predominant type of muscle fibres possessed.

Endurance fibres, nicknamed 'slow-twitch', are thin, slender and light, so they don't weigh the owner down. They are infused with a network of capillaries, ideal for delivering blood and, with it, oxygen and nutrients, while flushing the end-products of metabolism away to keep the muscle refreshed and able to keep going. They are populated by the energy powerhouses, mitochondria, rich in aerobic enzymes and the hardware needed to get a big return on breaking down carbohydrates, proteins and fats. The sprint type fibres, nicknamed 'fast-twitch', are big blocks of muscle, with strong fibrils to ensure high-tensile strength with high concentrations of anaerobic enzymes, ideal for emergency energy release.

Visualise your Sunday roast. As the chicken comes out of the oven, think about what the bird[xvi] was up to a few weeks before it was killed for your lunch. It would have been roaming around the yard, neck lurching forward

---

[xvi] It'll need to be free-range for the example to work, so treat yourself.

second-by-second with every step as it continuously searched the yard for a speck of feed, a grain, a bug or a worm. Their feet and leg muscles are slow-twitch, so they tire slowly.

But what if a fox were on the prowl? If he attacked, the chicken would be able to start running off, and would be able to get up to a fair old speed. But it wouldn't get up to speed quickly enough to avoid becoming the fox's dinner. This is where the pectoral muscles of the wing come into their own. The breast muscle, which, as you would know, are thick blocks of white muscle, are ready for the powerful emergency movements needed to avoid a fox attack[xvii].

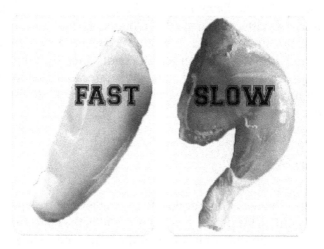

Figure 15. Chicken breast muscle (left) is packed with dynamic 'fast-twitch' fibres, whereas chicken leg muscle (right) is rich in 'slow-twitch' fibres. (Reproduced with permission from Colin Clegg copyrighted[4]).

[xvii] *Please see* Rocky II *for necessary application in training.*

An interesting feature about muscles is their plasticity. They are highly adaptable. If you stimulate them to adapt, they will respond to the stimulus and adapt to improve function that better equips them to deal with another similar stimulus. As we touched on in chapter two, if you give a muscle strength training, it becomes stronger. If you give it endurance training, it develops endurance.

The interesting feature is that, if you're an athlete predominantly composed of slow-twitch fibres (e.g. 85% of leg muscle fibres would be normal in a champion) and you do solely strength training, you will improve your strength, but not nearly as much as a fast-twitch creature could. At the same time, your endurance will deteriorate. The changes with strength training oppose endurance ability, i.e. the development of thicker, heavier blocks of muscle, with greater distance between the blood vessels and the mitochondria. The same is true in reverse for fast-twitch athletes who undertake endurance training. They improve their aerobic capacity and can develop their endurance, but not nearly as much as a slow-twitch creature.

SEE-SAW

Heptathletes are required to develop strength *and* dynamism for the shot-put at one end of the spectrum, and aerobic capability for 800m at the other end. While 800m is not quite at the extreme end of the spectrum of human endurance, as would be the case for the Tour de France – which involves four to six hours per day, nineteen days out of twenty-one – it is still considered an endurance event, classified as a middle-distance endurance variety.

As event duration increases, the proportional contribution from the two energy systems does not change in a linear fashion. Contributions from the aerobic system increase in a curvilinear manner. So the difference between the aerobic-anaerobic energetic contribution to an elite 400m is a forty to sixty ratio, while 800m is a sixty to forty ratio, even though the difference in event duration has effectively doubled from forty-five to a hundred and five seconds for men and from fifty to a hundred and twenty seconds for women. The shift of the aerobic system's input from forty to sixty, from 400m to 800m, makes it the predominant contributing system, and this alters the training needed to prepare optimally for a 400m versus an 800m race. For example, a 400m runner would typically run only 30 to 45 km per week, whereas an 800m runner would more likely run 70 to 100 km per week.

Herein lies the problem. An athlete like Jess will be providing a whole host of strength, high-force/dynamic, anaerobic and aerobic stimuli to her body. Ostensibly, these elements will oppose each other. Unlike the rowers, for whom the mixed stimuli will combine to help their abilities in their six-minute event, presentation of a shot-put stimulus will largely retard 800m performance, and vice versa. The strength-endurance continuum then acts as a see-saw. Add too much to one side and the other will be left dangling and will suffer. The challenge is to find the optimal balance.

Two major factors determined the tack that Jess would need to take. The first is easy to establish and requires only a little desk research. It is widely known by heptathlon coaches worldwide. Heptathlons tend to be won by runner-jumper-type athletes. Carolina Klüft, Jackie-Joyner Kersee and Eunice Barber are all incredible jumpers. To jump effectively, the athlete can normally run fast. What you rarely see is a world-class thrower winning in heptathlon unless their jumping is half decent. Right, that's the end of the heptathlon lesson! Next up, what is Jess good at and not so good at?

Well the good news is that she is a good runner-jumper. Super! When I asked Toni about her weaknesses, he gave me the story of her 2003 Junior World Championships performance. The event was in Canada and Jess led it at the end of day one, and indeed after event five. But then a poor javelin throw – some 18m behind the best throw – set Jess back from first to sixth. This was followed by a relatively mediocre 800m – some seven seconds behind the best time – which moved her up from sixth to fifth. The message was loud and clear from Toni. He would continue to work on her strengths, but her javelin and the 800m performances needed to improve. Previously, the 800m, as the 1,500m for decathletes famously had been, tended to be a bit of a torturous parade, which did little to affect the result. But now the best athletes were cottoning on to its potential as a weapon, and good athletes were getting decent points out of it.

My task was becoming clearer: it was to help develop the 800m. I knew it was a priority for Jess' event and I knew it was a priority for her physiological abilities. I also knew how to improve an 800m runner's physiology and performance. I had worked with more than thirty of the best ones in Britain, and in the years to come I would work with another thirty. You might think I was perfectly suited, but no, this is heptathlon!

With Redgrave, Macca and Mark watching over my shoulder, I began the search for knowledge in this area by asking Toni what they did. It is somewhat one-dimensional simply to extract the running-specific training and share that. The full training programme gives the broader picture and the full interpretation.

This typically includes the mainstay of '200m type' (which typically involves repetitions of similar distances, such as 150m, 250m and 300m) training. This session is a cornerstone for 200m performance development, but it also underpins 800m anaerobic development.

Figure 16. Discussing potential training sessions and plans for competition with Toni Minichiello and Jessica Ennis-Hill in early 2016.

Heptathletes obviously do hurdle training, which is also given running credit. They also do some short, sharp sprints. Hurdles and sprints obviously count as high-speed running training, but they are also high-impact, so this needs to be accounted for and weighed up as part of the heptathlon recipe.

The other major type of session would be 200-300m reps, but with short recoveries aimed at developing the 800m. A typical session would comprise three sets of 2 x 200m, with thirty seconds between 200m and six minutes between sets. This trains the tolerance of hard anaerobic efforts and also trains the endurance of 800m. Most heptathletes would run this type of session at just a slightly faster pace than their 800m running speed.

Toni was currently using these sessions, so this was the starting point. As I had seen previously with the rowers and middle-distance runners, I needed to start with what the coach and athlete were currently using. Proposing something completely alien, or something that doesn't fit, is never a successful approach to intervening with these high-performing specimens. However, that doesn't mean you should completely neglect what is documented; what is published; what currently sits in the body of knowledge.

No items found

As a good scientist, I scurried off, ready to do my research, and undertook a search of the major databases. I used the scientific leviathan of a search engine, PubMed, as a starting point and posed some search terms for its delectation.

"PubMed, give me everything you've got for '800m training heptathlon'."

It didn't take long to return: "No items found."

'Okay, let's broaden this a little,' I thought. 'How about "800m heptathlon"?' The same thing happened again. "No items found."

My frown and bottom lip came out to play, and I thought. 'Really? Is there truly nothing worth publishing on 800m running for heptathletes?'

In a desperate attempt to find something relevant, from which I could forage for clues regarding the direction I would need to take, I simply entered, "Heptathlon." You've guessed it. "No items found."

In fact, I'm sure PubMed did a little evil pantomime laugh at my foolishness. By 2016, PubMed would return only four articles related to this topic. Three were systematic reviews of injuries at International Association of Athletics Federations (IAAF) competitions, one of which examined masters (veteran) athletes. The other study explored antioxidant damage after heptathlons. The study concluded that it is a stressful event!

My research continued in the background, searching coaching journals, sporting libraries, multisport texts and so on. In just a few steps I was firmly back in the territory of coach knowhow; the same material, educational resources and personal accounts Toni would have been schooled in. This was not a bad place to be, but it meant there was no other reference point for Toni or me to use, reflect on and brainstorm with.

The implications were now apparent. To start with, we had Toni's training, but we had no other observations or records of good athletes. We had a couple of options at this stage. We could either begin to unpick the training sessions or we would need to learn more about Jess' physiological makeup. Jess wasn't ready for a laboratory test, despite me cajoling her. One look at the almighty treadmill and she baulked.

With my attention on understanding the entirety of the programme before we got specific, we began to work through Toni's training. We met up every couple of months and started to debate the sessions, discuss them, rip them apart and play about with them, but, essentially, we were considering the sessions he had always used. This early phase of our work saw little actual change. Toni asked a lot of questions about what was going on during one session or another, or what would happen if we did this session on this or that day.

We started to reach some firm conclusions about certain training sessions and, most importantly, how these sessions would progress over time. I was keen for some of the changes to start manifesting themselves in the training programme, but it was for Toni to make the decisions and introduce any modifications.

During the 2005 season, he made some minor tweaks. He simply didn't see the need to change a great deal at this stage. I think he was warming things up for the following season, as Jess began to question a few of the ideas Toni had shared.

"Have these new training sessions come from you, Steve?" she asked suspiciously. This was a question she would continue to ask throughout her career whenever a new session reared its head.

Toni was mounting a plan for the 2006 Commonwealth Games, which were held in February that year, and began to introduce our ideas in the off-season of 2005 through to Christmas. He then needed to switch the training to a taper unusually early. This was an opportunity for some further support. I joined several of their track sessions, which normally took place with just the two of them rattling around the English Institute of Sport's indoor track in Sheffield.

My job was to take some measurements and give them feedback about whether the session was doing what Toni and I had intended. In many ways, it was, but in others it wasn't. I had observed, for example, that the intense sessions – the repetitions of 200m to 300m – weren't intense enough. We would later go on to change this, but this wasn't the time. However, Toni generally did a lot of tidying up of Jess' sessions and we had to let these small changes sink in.

My other job was to think about the overall recovery process for the heptathlon. Once again, there is no reference material for recovery from these multiple events. We had to build this from scratch and think through the logic of the demands of the events, the windows of opportunity to intervene and any subsequent events.

There are ample opportunities to intervene after the second event, the high jump; after the 200m at the end of day one; after the first event on day two, the long jump; then briefly after the javelin. These are the only opportunities to replenish with fluids, energy, protein, antioxidants and supplements, and to get some rest.

However, you can't recommend that an athlete shovels a load of food and drink into her mouth just before she is about to project herself over a high bar or a long way into a sandpit. Equally, high-profile athletes will probably do some media work, the athlete will need to be transported to and from the accommodation, and the exact facilities needed might not be available in the bowels of the stadium. The recovery plan was designed to be specific, but also pragmatic and flexible.

Jess made real progress in her performance at the Melbourne Commonwealths. She had an outstanding high jump, enough to have won the gold in the single-discipline competition, and made a significant step forward in the 800m, taking her personal best from 2:17.03 to 2:12.66. In heptathlon money, the 4.37 second improvement in 800m equates to sixty-two points. The difference between Jess' third place and Jessica Zelinka's fourth was fifty-six points! We had made up some good ground. A four-second personal best was a big step forward.

Toni was starting to assemble a crack team to support Jess, known affectionately in the press as Team Jennis. Paul Brice, the biomechanist; Alison Rose, the physiotherapist; Derry Suter, the massage therapist; Mick Hill, the javelin legend and now a coach, were among the early signings, and I was of course part of Team Jennis myself. Later on, Toni would take on Mark Ellison to tackle

Jess' nutrition, Pete Lindsay to work on her psychology and Rob Johnson as a more local physiotherapist.

When I first started working with Toni, several of us were doubling up in some of the roles. For example, early on I advised generally on nutrition, but as we progressed we saw a need for specialised input. For me, this meant any influence I had in that area had to be handed over to someone who could offer a greater contribution. Any ideas I had needed to be fed in to Mark, trusting in his greater nutritional knowledge to take it forward.

Figure 17. Team Jennis (or some of them, at least, left to right) Derry Suter (soft tissue therapist), Toni Minichiello (coach), Alison Rose (physiotherapist), Rob Johnson (physiotherapist), Dr Paul Brice (biomechanist), Jessica Ennis-Hill (no introduction necessary), me and Mick Hill (javelin coach), celebrating Jess' 2009 World Championships title.

The expansion of Team Jennis also meant we had new partnerships and people to agree with. When we were working on Jess' inspiratory muscle training, I remember Alison Rose having an idea about it not being performed passively; sitting down, for example. She suggested that Jess did her breathing training while performing a leg exercise, such as high knee lift, while standing or sitting on a Swiss Ball.

I didn't think much of the idea at the time as I just couldn't see the connection, but I dropped Professor Alison McConnell (inventor of the POWERbreathe training device) an email to see what she thought. Lo and behold, she said that the idea was spot on due to the interconnections between breathing movement and limb actions. Hat tip to Alison Rose for thinking holistically and laterally! This just shows the strength of a group of people with a team's worth of ideas, where expert discipline specialisms add the real depth of quality.

Warming up endurance

By this time, Jess was warming to the work we were doing. It had given her results, even though some of it was more painful. She was certainly engaged and liked to know what was going on and why. This was mainly because she didn't want to do anything unless she needed to, but knowing also gave her more confidence.

Despite being more interested and accepting, Jess was nonplussed about being put through a treadmill test. Ideally, we would have profiled her throughout her career, but once she had agreed to it, she made it clear that a repeat performance was unlikely once we had derived the major information from it. We knew we needed to maximise the opportunity. The test took place in November 2006.

What did we learn? The headline news was that her VO_2max was very good. From my prediction models, I suggested that her VO_2max indicated that she would be able to run 800m in under 2:10, faster than she had performed at the time, which showed her potential. So we knew that she had a good engine. This was great news. It's always easier to get potential out of the capacity rather than having to develop the capacity.

However, her base level of fitness, her ability to produce less, or clear away more lactate, was low. As we had seen with Hayley Tullett and the 1,500m, lactate clearance might not be a key determining factor in performance, either for the 800m or for heptathlon, but as we saw with the men's 8+ it can hold the whole system back if you have a poor lactate clearance capability. It had to improve, but how could we do it?

Figure 18. Jessica Ennis-Hill undertaking a physiological test with me in 2009.

To develop lactate clearance, a marathoner would do a session such as 2 x 20 minutes at threshold. A 10km runner would do the same, or maybe 3 x 10 minutes at threshold. Hayley Tullett and other 1,500m types would do 5 x 5-6 mins, while 400m runners would do 4 x 4 mins. For Jess, this session wasn't only alien; there was no room for it. She couldn't do an extra running session as this would have further ramifications. She couldn't replace another session with it, because that would deprioritise[xviii] another event.

I was certain we would have to find a way, but to make it fit again we had to adapt: adapt our knowledge, adapt our thinking and adapt our interventions. As Toni said: "We need to do something new and innovative; something no one has ever done before."

So I mapped out everything Jess was doing; every throw, hop and stride. The running sessions were generally sound, with only a little room for manoeuvre. We had accepted that.

Then I sat back and scrutinised the training. Jess did ten sessions each week. For each of these, she warmed up by jogging for approximately 800m, which totalled 8,000m per week. Jess would tend to plod the 800m in about six minutes for her warm-up, which probably did a decent job of warming everything up. Six minutes for 800m equals 9.5 km/hr.

So I asked myself the question: could Jess improve base fitness if she ran her warm-ups at the pace we would normally prescribe for threshold training? Any physiologist in the world would probably tell you, "No, almost certainly nothing would happen. The stimulus is just too small and insignificant compared with everything else she's doing. Go back to physiology school, Ingham." I knew I was trying to be adaptable beyond the reasonable level of acceptable adaptability with this idea.

[xviii] *Hopefully you are now getting a sense of the plate-spinning exercise of heptathlon training.*

But what if we thought about it in a different way? Those six-minute sessions would add up across the week to sixty minutes per week. So that's sixty minutes multiplied by at least forty training weeks per year. The overall magnitude of the stimulus might be small on a day-to-day basis, but gently and slowly it would increase the base fitness without damaging other areas. Also, pragmatically, what other choice did we have other than to try it? Wasn't this about doing something innovative? Didn't we want to peak for 2012, rather than tomorrow? Wasn't it my job to put forward ideas that helped the whole event; not just to make Jess a half decent 800m runner and a rubbish heptathlete?

I presented the idea to Toni and Jess in this order:

- Base fitness needed to improve. They agreed.

- We shouldn't over-focus on this area right away for fear of disrupting other areas. They agreed.

- It needed to be addressed gradually over two to three years. They accepted this.

- Jess was already doing six-minute warm-ups. Yep.

- So why don't we tweak up the pace of them to see if we can nudge this area of fitness up? Yep, that makes sense.

Their first reaction was, "Is that really going to work?"

I had to be honest. "I have absolutely no idea!" I said. Then, in a 'secret squirrel' moment, I added, "But I can be pretty sure no one has ever thought of doing this before." Just to be sure, I tacked on, "And it's unlikely to do any harm."

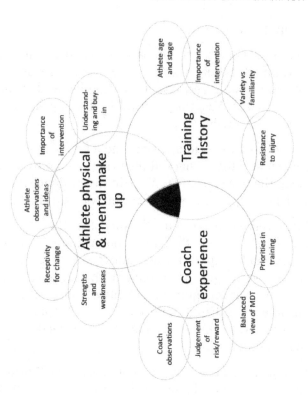

Figure 19. An example of a number of high-level factors a coach and multidisciplinary team need to weigh up before reaching the central fulcrum of a training intervention decision.

They went for it.

I gave Jess a new warm-up pace and off she went. Yes, the revamped 800m warm-ups were a bit on the uncomfortable side. Yes, Jess did let me know that they weren't pleasant. Yes, they started to develop her base fitness ever so slowly.

We collected the evidence over the months and years, reviewed it as a team, checked the progress and tweaked it to keep the pace effective. We held the evidence in our hands. The crazy creative intervention was working!

The intervention began to increase the rate of adaptation just enough that it was outstripping what we would normally expect to see, but it wasn't changing wildly, as we might have seen with full focus and attention. It was nudging her abilities along at a rate appropriate for a heptathlete who needs to perform a good 800m and, for her physiology to do so, needed good metabolic fitness.

Carefully paced warm-ups do not maketh the World or Olympic heptathlon champion, but this was a critical example where, if the eventual intervention was held up right next to a textbook of human exercise physiology – held against known methods and subjected to academic scrutiny – it would get laughed out of town.

A further example of the development of ideas was how Jess warmed up for the 800m. In the early 2000s, physiologists Professor Andy Jones and Dr Mark Burnley were developing insight into the priming effects of warm-up, seeing that, at the start of exercise, the body can initiate its systems at a greater rate if primed in the right way.

I had taken this information to several coaches to explore whether they would like to begin to use it for their athletes. The warm-up is often quite a delicate area, and neither coaches nor athletes want to be experimenting in any way that is excessively taxing for the athletes' bodies just before the delivery of a performance. Anything that asks them to overexert in the lead-up to a performance is highly questionable.

Andy and Mark were developing an insight into priming the body, allowing time for the negative consequences of the effort to be recovered, but they were also observing that there was a residual priming effect, whereby the muscles were warmer and had a greater amount of oxygen in them for the subsequent effort. This led to the athlete being able to consume more oxygen and fatigue less during all-out performances.

When we undertook some research, we found we could replicate this effect on elite-standard 800m runners, although it would be a further three years before we could successfully convince, introduce and refine this method for Jess. The end-product not only featured the 'priming warm-up' but also incorporated inputs from Alison Rose, Jess' physiotherapist, Mark Ellison, Jess' nutritionist, and Pete Lindsay, Jess' psychologist, to ensure that it knitted together neatly.

It was part of a project that focused on ensuring Jess warmed up and recovered from each of her individual events as optimally as possible, so that her overall heptathlon performance improved. By the time heptathletes finish the first day, they're quite tired. By the time the fifth and sixth events roll around, they're even more tired. While it's common to see a heptathlete acquire a personal best in an 800m event, especially if they're chasing a medal, it's rare to see a heptathlete produce an 800m they would ordinarily be capable of doing without the preceding six events.

Therefore, the project aimed to deliver Jess to the 800m as recovered as possible and, more importantly, better recovered than her opposition. Therefore, we were plotting a way of acquiring points in the heptathlon without having to alter too much of the training programme. The recovery plan gave her points for free!

Ideas such as warm-up priming and the whole event recovery plan are complex and take time to fully develop. From the initial research suggestion, the priming warm-up was nine years in the making before it was ready for Jess to use in competition.

Just as a reminder, the essence of this book is about the craft of applying science to talented athletes in the real world. Given the set of circumstances we faced, the demands of the event, the psychology of athlete and coach, the individual physiology, the openness, the practicalities, the need to act, to trial, to test and report back, this was a reasonable idea, and one that we found to be effective. Judging an idea without these contextual inputs is short-sighted, narrowminded and ignorant to the real world.

Indeed, a host of ideas, interventions and changes were discussed, agreed upon, trialled, introduced and became standard practice for Jess. Equally, a host of ideas and interventions were rejected at any given stage of discussion, agreement, trialling and introduction.

THAT OLYMPICS THING

Let's fast forward six years. Within a week of the Games starting, and just a few weeks from finding out whether Toni's ambition of peaking Jess for the 2012 Games would materialise, I visited them both in Sheffield to dot the i's and cross the t's to give her my laminated guide cards for the recovery planner and 800m execution plan.

I was struck by how tense Jess looked. She was bearing all the pressure, expectation and hope of Team GB on her shoulders and her normal laidback approach was starting to take the strain (as yours would if a 1,000m² portrait of you had been painted next to the Heathrow runway to greet every Olympian, coach, official and spectator to London).

I am sure you are expecting me to tell you that I calmly and confidently cracked a joke and made her feel great about herself, but I didn't. I saw the nerves and yet, despite all my experience, I just started to talk, in fact moan, about how a journalist had misquoted me about altitude and how it had momentarily landed me in the bad books of a performance director. What a numpty! I knew I was doing it at the time and tried to snap out of it, but I made a total mess of it. Again, what a numpty! I wished her farewell and told her to follow the plan. There were points to be won if she did. Slightly less of a numpty!

I took the train down to London St. Pancras that evening and then on to the Olympic Park for a meeting. I was shaking my head all the way down about how clumsy I had been. As the stadium came into view with the iconic ArcelorMittal Orbit looming large at me through the train window, my goosebumps flared and I reflected on the long journey that Jess and so many athletes had taken to arrive at this point; just before the curtain lifts and the all-important performance takes place.

I knew at that moment that if Jess were to execute her events to her full ability she would win, but knowing this did nothing to diminish the sense of the unknown and the great uncertainty. A wave of anticipatory excitement and an urge to know the result came over me. Little did I know then that, as her name was announced for the 100m hurdles, as the visible tension in her eyes began to lift as she took one big, soothing breath in, her unmistakable and truly genuine smile would ignite the stadium. From clearing the first hurdle to devastating the field with a hurdles time equivalent to the individual event gold medal winning time from the 2008 Olympics (12.54s) through to the last tumultuous lurch for the 800m finish line, she would sail through the competition with all the relief of a whole nation.

Figure 20. Jessica Ennis-Hill during the 800m, on her way to becoming Olympic champion at the London 2012 Olympics.

As these inspirational thoughts of what could be wafted over me, some fellow travellers' chatter punctured my haze of daydreaming with a killer blow, revealing that my hope and ambition to help talented athletes become the best in the world was not a view that was shared by everyone.

"That Olympics thing, I think it's going to be so shit. A complete waste of time! That new Westfield Centre, though, that's more like it."

I needed all my Jess-induced adaptability to tolerate this alien view of my beloved sport. I hope, as many Brits did in the summer of 2012, that they watched Jess a few weeks later and were moved and inspired to recognise the enormous pride the Games gifted the host country.

THE EIGHTH EVENT

Between 2012, Jess decided to throw a new physiological challenge our way: a pregnancy! Her son, Reggie, was born in July 2014, meaning that Team Jennis was put on notice of working on 'the eighth event' from January 2014. Our brief was clear: "Can Jess come back from giving birth and still be competitive?"

Toni pulled the team together and we each shared some thoughts about what happens to the body during pregnancy and after the birth, and how to load again in the weeks and months that followed. We were all awash with details about hormonal changes, abdominal strength, laxity of connective tissue and so on. Toni was pressing each of us about how long Jess was likely to be able to train into her pregnancy and how soon after the birth she could recommence, but we were all entering the unknown.

Once again, we were confronted with "No Items Found" search data, as there is no formula for going from Olympian heptathlete to the Herculean event of parturition and back to Olympian. When working with such unknowns, you must usher in a whole new level of watchful monitoring. Toni needed to stay very aware, plan carefully, be adaptable and manage his own expectations. Jess needed to become even more in tune with her body and feed back her observations, no matter how subtle, to the team.

As our team meeting delved into the detail, it was the sage words of Dr Richard Higgins that lifted us up to the most important perspective of all.

He said: "Look, folks, we can discuss this all we like, but the most important thing in all of this is that Jess has a happy and healthy birth!"

Richard's words kept ringing in my ears throughout Jess' comeback. As for so many people upon becoming a parent, her life perspective changed. For her training, it meant we were given a different brief; this time from Jess rather than Toni. Her focus had changed and she wanted us to construct a programme so that her training would be done and dusted in the morning. There was a simple reason for this shift: to ensure that she had the rest of the day with Reggie.

As Jess put it: "It used to be twice a day, six days a week. Now I just do three or four hours in the morning and then have the afternoon at home."

From a physiological perspective, this was a step backward, as the extensive rest and recovery we had built in to allow for maximum adaptation between training sessions was being squeezed.

But Jess was very clear: "He is my priority, and everything has to fit around life with him."

Jess began her comeback in earnest over the winter of 2014-15. Her overall goal was a medal at a global competition, whether it be Worlds or Olympics. At first, she found that her usual rigid structures had softened somewhat, leading to a lack of strength, efficiency and co-ordination. However, Jess found the longer training sessions easier, due to the boost in blood volume that coincides with hosting a baby.

It took time for Jess, Toni and the team to steadily load the training up again. She made steady strides forward, helped by Toni's inspired rebranding switch in mentality, from introducing PPPBs (post-pregnancy personal bests) to saying, "Forget what you've done in the past. Our expectations need to start all over again." Overall, the emphasis swung to quality over quantity.

When Jess entered the 2015 Beijing World Athletics Championships in Beijing, she was hoping for a reasonable performance, a possible medal. It was mainly viewed as a tester to see if she could be competitive for the 2016 Rio Olympics. She only went and won the thing!

This is interesting for a few reasons. First, it was a win! It was an incredible achievement to return within fifteen months of giving birth and take a global track and field title. She was only the third person to have won a world athletics title after having given birth. Second, her score was substantially lower than her best points tally (6,669 versus 6,955), though at a similar level to the winning score from the 2013 Worlds (6,586 by Hanna Melnychenko). Third, given that her overall comeback goal was to win a global medal of any colour, you would have to say it was mission accomplished.

In 2011, Jess was beaten to the world title by Tatyana Chernova. In November 2016, it was announced – in what was becoming a frustratingly recurring theme – that Chernova was to be stripped of her gold medal, meaning that Jess became, retrospectively, three-time World champion. This affair sets off a storm of emotions, blending vulgar injustice with patient fairness.

At the time of the loss, Jess was extremely disappointed not to have won. We all were. But the loss was perhaps the best thing for her in 2012. It gave her unremitting focus and determination throughout the season of the home games. Paradoxically, this lack of success had fuelled her spirit. In 2016, she had higher priorities, ones that were loftier and more noble than heptathlon – her son.

At a Supporting Champions: Conference in 2017, Toni reflected, "I often wonder if there was more I could have done to convert the 2015 Worlds win into a 2016 Olympic win." He (and we) will never know.

But just as with all sport, there are so many permutations and eventualities that you can never be sure that A+B=C. Pinning analysis or opinion on one set of circumstances to explain a medal outcome is a foolish business. What we do know is that Jess entered 2016 as a keen favourite for the title. The big tussle was meant to take place between Jess and fellow Brit, Katarina Johnson-Thompson, but nobody anticipated the extraordinary talent of Nafi Thiam from Belgium, who came racing through with five personal bests. Thiam's score was beatable, and if Jess had achieved a better shot-put and long jump, she would have taken the title, but boy did she fight for it.

While aspiring to retain her Olympic title, she chose to do so as a mother first. As such, she showed us that life doesn't need to be put to one side to win at all costs.

In the words of journalist Oliver Holt: "Jessica Ennis-Hill burnished her reputation more in defeat last night than in many a victory. A great example of a human being with class."

In terms of words used to describe Jess, I can't improve on that!

From 2005 onwards, Toni laid down his inspiring challenge and I took it up with all the excited anticipation of trying to nail jelly to the wall. Working with Toni, Jess and everyone else within Team Jennis showed me the need:

- To remain adaptable in applying systematic observations to complex situations.

- To take a leap of faith with the information we had in play, using first the principles of training and recovery; variety in training; strengths and weaknesses; and of course human physiology and psychology. Trial and error would act as our guide rope as we climbed through the uncertain ups and downs of this unusual mountain.

- To be prepared to park the statement, "The trouble is…" and ask the question, "What if…?" and "Yes, and…" to ensure that our ideas could turn into innovations.

CHAPTER 5: JOURNEYING THROUGH KOCS

"One's philosophy is not best expressed in words; it is expressed in the choices one makes... and the choices we make are ultimately our responsibility."

Eleanor Roosevelt

2 x 7

As Jess Ennis was on the rise to becoming arguably the best female athlete in the world, her breakthrough moment was the 2006 Commonwealth Games. The event had been won relatively comfortably by Kelly Sotherton, though it wasn't her strongest performance. I had watched the competition as a fan and as Jess' physiologist, willing her on through her seven events.

When you're in someone's support team, you obviously want them to win. So, effectively, I was pleased that England registered a gold and a bronze medal, but I would have been more than happy for Jess to have beaten Kelly to take the top spot. Your loyalties are normally aligned to where your duties lie. You are invested in the process, but there is also an inherent respect for athletes as they put themselves forward to compete. Like them or loathe them, you want to see them achieve their goals if you are involved.

Fast forward to the 2007 World Athletics Championships in Osaka, Japan, and my loyalties were firmly divided. I was supporting Kelly *and* Jess, and I wanted to see them both succeed, but if I had been coerced to answer the question of whom I wanted to succeed more, I wouldn't have been able to. I will attempt to explain what led me to this position.

Kelly Sotherton burst onto the athletics scene in 2004, taking bronze at the Athens Olympic Games. If you look at her record, you can see that she should have been further ahead in the years preceding this. She had been an outstandingly talented junior across a range of disciplines, but had enjoyed herself a bit too much to make serious headway in her early twenties, and never really nailed a full heptathlon.

Charles van Comenee, her mercurial Dutch coach, had steered Denise Lewis to Olympic gold in Sydney and had taken Kelly on in 2003. His determination, drive and relentless pursuit of comprehensive standards had rubbed off on Kelly. In 2004, she entered the Olympic Stadium as an unknown quantity, but both Charles and Kelly knew from the season's performances that a medal was probable.

After the 2004 games, Max Jones, the performance director of UK Athletics, retired and the vacancy came up. Charles was considered for the role, but it didn't happen and he decided to head home to his native Netherlands to lead the country's athletes to their most successful athletics meeting at the Beijing 2008 Olympics. Kelly had lost her coach just as her career was taking off. On top of this, she faced the minor problem of having one of the world's greatest ever athletes, Carolina Klüft, ascending the heptathlon world stage.

Being the bold and brazen soul she is, Kelly decided that no one in the UK could match Charles. If she couldn't have the best in a single coach, she would take the best of a collection of coaches and assume the responsibility of pulling it all together. For the 100m hurdles she employed the vastly experienced Graham Knight; for her high jump she employed the flamboyant Fayyaz Ahmed; for her shot-put she employed the legendary Shaun Pickering; for long jump she employed the cool wisdom of Aston Moore; for the javelin she employed Mick Hill, then John Trower and then Mike McNeill. The javelin was her problem event.

Kelly took control of the running. She used the sessions that had served her well throughout 2003 and 2004, and replicated them. This worked in her favour only because she was in control, but the results were patchy. Kelly's strength was her running. She regularly beat her training partner and compatriot, Denise Lewis, and became one of the strongest runners in world heptathlon.

Despite Charles labelling her a wimp after the Athens 800m, where she was in contention for the silver medal but got bronze, Kelly was one of the first women to exploit 800m for its full medal potential. Previously in heptathlon, there had been a couple of women who were good at 800m but these tended to be women who did not affect the medal placings. Now Kelly was showing that potential medallists needed to be better at 800m or they would be punished in the eventual standings.

Despite achieving personal bests in the 800m and 200m between 2004 and 2007, the fact that she wasn't making consistent improvements was noticeable. Kelly and lead coach Aston began to notice this and started to discuss options. That was when they came to me.

FAST START

I met Kelly for the first time in February 2007. I had come to the Athletics Centre at Loughborough early for our meeting and Kelly was training her javelin technique by throwing a small ball against a wall. She was working with her biomechanist, Dr Paul Brice, who I knew well.

Kelly is certainly known for being brash, and that is how she tends to behave when you first meet her. She opened with "What are you doing here? You're not supposed to be here for another thirty minutes!"

Kelly would often find some sort of quip. It was her loving way of sussing you out. She seemed to thrive on this approach and you could see Paul and Aston reeling back, shaking their heads and covering for her with something like, "Don't worry, she's like this with everyone."

Kelly and I sat together in the upstairs area of the centre, where she had some specific questions about the use of a supplement, sodium citrate, which is known to help alleviate the feelings of 'acid burn' during the high-intensity efforts of training or racing. Having worked with the supplement during my time with the rowers, middle-distance runners and 400m runners, I outlined the potential options for her.

She then rapidly moved on to questions she had about training for the 800m. I got a couple answers in before taking a breath. She then accelerated again, asking about the pros and cons of certain types of training session in relation to the type of physiology she had. I said I couldn't possibly know! I was answering and deflecting her questions as best as I could, but having worked with similar athletes on an in-depth basis, I needed to fully understand how she and her team were preparing before I could begin to provide insight and judgement about the effectiveness of their methods. The problem was, I wasn't being given the chance.

However, Kelly was beginning to gain momentum and the questioning continued at rapid-fire speed. She moved on to pacing, then breathing, then strength and conditioning for the 800m.

She asked: "What do you know about 200m training? How do you get the most out of 200m training and 800m training at the same time? What are the best sessions to peak for 200m performance?"

The questions were coming thick and fast. Kelly clearly needed assistance and the answers I gave did nothing to deter her.

"What do you know about sprint training for 100m hurdles?" she continued.

These questions were straying far beyond a normal consultation. I was relying as much on my physiological knowledge as I was on my own understanding and experience of athletics as an avid reader, albeit as a one-time (slow) sprinter. As a runner, I had adopted some quite different approaches to training, using as much of my physiological knowledge as I could to help me go as fast as possible. However, I also knew that you needed to choose your parents very carefully when you were making the decision to become a champion, so no matter how much I knew and applied to my training, I wasn't going to be making an Olympic final any day soon!

In the space of a single hour, we had covered twenty topics, talking about training, 800m running, 200m running, sprinting, recovery, rest, physiology, nutrition, mechanics and psychology, to name but a few. Kelly was clearly open-minded enough to want to hear a range of opinions. It was obviously her way of evaluating me; of assessing whether I had a potential role to play in her development. Given that she was so hungry for success, it was clear that she was prepared to consider all the options of who might support her and how they would do so.

This was the first time in my career I had experienced such a swift rate of questioning. The extent of topics covered went well beyond the normal range regarding physiological responses. She was stretching the questions to a much deeper level. There was also a pace and intensity to her questioning that hinted at the extent of her ambition. I had never encountered anyone so willing to accelerate the conversation to the point of decision-making and action so quickly.

I remember thinking, 'At this rate, she'll be asking me to coach her in the javelin by the end of the conversation!"

Luckily, she didn't.

ACTION

We left the discussion holding the outcomes typical of a six- to twelve-month-old scientist-athlete working relationship. We had agreed that I would observe her training, and to undertake a physiological profiling test. We had also agreed to sit down again and discuss her training methods.

To my surprise, she was on the treadmill undertaking a laboratory test within a week. Her stance was, "Let's do it."

She needed to see results and she wanted good results that season. The window between March and May is the last real opportunity to make substantial changes that will influence performance in time for the summer.

I processed the results. She had a good engine, but it wasn't good enough. She had good lactate clearance, but it wasn't good enough. She had good economy, and it was certainly good enough. In fact, it was exceptional, especially for an anaerobic-type athlete.

It is always good to be in a position to tell an athlete why they are good at what they do. All too commonly, sport scientists tell athletes what they need to improve on, how they can go faster or that their abilities aren't good enough. I think it should be the other way around. A sports scientist needs to lead with the positives. What are the strengths of the athlete? What are the components that explain why they are excellent? An athlete needs to hear this.

They already know they are good, otherwise they wouldn't be at the top of their sport, but they might not know *why* they are good. I think sports scientists play on the negatives because they feel they want to be involved and depended upon! Focusing solely on the negatives is a mistake as it puts up a barrier on the part of the recipient.

At the very least, the 'feedback burger' should come out. You know the one, where you give a very positive start, no matter how tenuous, followed by some meaty negatives, completed by another positive to round things off. The bread offers the buffer, but you can always taste the meat!

Kelly absolutely needed to hear her strengths. Although she wasn't there just to hear what made her run fast, she wanted to know what the deficiency was and made this point very clearly. Almost before we had covered the full debrief of her physiological profile, she asked what she needed to change. The focus was on the training and the process.

She needed to know what her prospects were. In true Kelly Sotherton style, she wanted to know whether her fitness was good enough at that moment in time, and whether she would need to explore other avenues and answers to explain the decline in her performances. She wanted to know whether, if her fitness wasn't good enough right then, it could be recouped in time for the 2007 World Championships. I gave Kelly the information and she processed it at an incredible rate. Her ability to focus on action was hugely impressive; some might even say gung-ho! However, she also sensed an opportunity and discerned that, if she didn't act quickly, she would miss it.

I was in the middle of moving seamlessly from sharing information about her physiological profile into my recommendations when she stepped on the gas and moved the conversation far beyond anything I had experienced before.

Typically, a physiologist provides recommendations for a coach or an athlete. Normally, you draft sessions for athletes. You provide them with an outline session they can understand, get their heads round and bring to life for themselves. For a sports scientist, the line at which you stop, and the athlete and coach take over, is normally at the proposal of new ideas based on their individual, bespoke physiology, psychology, dietary habits, movement patterns, injury, or whatever else it might be. Under most circumstances, the athlete and coach will then pick the baton up from you, the expert in your specialist field, and will add their own expertise, context and understanding of what makes them tick. Then they will integrate, reject or park your suggestion.

It is not uncommon for a sports scientist to straddle this line by working with the athlete and coach in the training environment to ensure the baton is passed on securely. This could take the form of the sports scientist processing his or her ideas, presenting them to the coach and athlete, and then accompanying them to the training sessions for a more detailed application of the recommendations to the training they are currently doing. This might involve further measurements, observations and discussion or negotiation. This step bridges, with greater confidence, the cold, scientific observations to the fuzzy, complicated and holistic set of circumstances the unique athlete and coach pairings face.

Rarely, and probably wisely so, does the athlete and coach pairing take your recommendations and simply implant them into the training programme without first evaluating them. If this does occur, it tends to be carried out by relatively inexperienced personnel. The recommendation must be integrated, but occasionally a coach or athlete will take the recommendation and simply add it on top of what they are already doing. This is a big mistake because managing training load is a critical aspect of an athlete's progression. If both the recommendation *and* the way it is integrated are not fully considered, disaster will usually follow.

The question that should be asked is, what is going to be taken out to make room for this recommendation to be effective? A coach needs to know what this new session will look and feel like. It's a bit like a time-management recommendation of only doing an hour of emails per day. It goes without saying that this recommendation supersedes the previous practice of, say, three hours of emails, rather than being added to it. A training programme can't simply absorb more and more without re-evaluating the existing content.

Occasionally, the coach will simply take the recommendation at face value and swap it in for another session. Again, this is a mistake because it doesn't show the full perspective or appreciate the priority of each of the different sessions in use. Simply swapping in and out based on a scientist's recommendation is not necessarily a recipe for disaster, but it can certainly create confusion. In this circumstance, handing recommendations over to the coach is the first step, but it is equally important that a scientist ensures that the coach takes ownership of the session and applies it to their specific context if the recommendations are to have any chance of succeeding.

AUTHORSHIP

Kelly was familiar with sport science and medicine support. She regularly received treatment from experienced therapists and medics. She knew how it worked. She was prepared to do whatever was necessary to get back up to her previous level and extend to new heights. In a moment that I won't forget in a hurry, while I was halfway through verbalising and drafting some suggestions for her programme, she thoroughly socked it to me with the intensity of a stranger proposing marriage. In a single sentence, she moved the line beyond the point where I had stepped previously. That line had been my professional boundary. That line had, in many ways, protected me as a service provider.

She asked me, "Will you write my programme?"

The question was much more loaded than it might sound. It's one thing to write down the training sessions, or even to draft your recommendations, but to author a training programme is an entirely different prospect. The translational nature of presenting ideas, passing them over and leaving the athlete and coach to their own devices is a much safer place to remain. Writing the training programme is to truly take ownership, responsibility and accountability for its effectiveness.

Ownership forces you to recognise that others will view it as *your* programme. The responsibility causes you to feel the weight of that ownership, aware that you must be extremely careful about what you create for the programme. The sense of accountability should infuse through you, so that when you put pen to paper and then publish the programme by forwarding it to the athlete for their digestion, and when they begin to execute the programme with every sinew of their efforts, you need to be able to explain why they're doing it and to analyse the results that follow.

So my response was one of clarification: "What do you mean, write your programme?"

"I want to know if you'll manage my running programme?"

"But I'm a physiologist!"

"Well, you know what you're doing. The suggestions sound great. You're basically writing the sessions anyway. So why don't you do it?"

"Okay, maybe. I'd have to think about it."

I began to feel myself getting drawn along. As I did so, I started to form some conditions. If I was to write the programme, I wanted to see it implemented the way I wanted it to be. And if I was going to oversee the training sessions, I would want to take every step available to me to ensure that I knew, with as much confidence as possible, that the athlete was moving in the right direction.

The entirely obvious outcome of writing a training session for somebody else is that they will end up doing your training session, trusting that you have given that session a great deal of thought as it is their efforts you're playing around with. I explained this to Kelly and she bluntly gave me the clearance to go ahead.

"Yep. Do what you need to do!"

As soon as the gauntlet was thrown down, I found myself reaching out to take it on.

I don't know how other coaches feel when they're writing a training programme. I imagine former athletes (and by that, I mean athletes in full-time training and having competed at a decent level) might find this relatively easy because they might replicate and adapt what they have previously done. I know some former Olympic gold medal winning athletes who struggle to prescribe anything other than the training programme *they* performed.

Perhaps the achievement level provides sheer certainty about the methods. This is completely logical. If success came from a process, the process is validated. However, what it doesn't do is provide certainty that that template of training provides the optimum stimulus for another human being with a different set of genes, with a different training background and a different mindset. At worst, this can lead to a copy and paste application of the same training blueprint. It might work, but often it fails to hit the spot.

For the rest of us failed wannabe athletes, and for anyone with sufficient conscientiousness, interest and open-mindedness to get it right, the prospect of writing a training programme for an aspiring Olympic medallist is daunting. When I took on the challenge of writing Kelly's training programme, I imagine the process was like that of an artist, painter, storytelling author or landscape gardener about to begin a new project.

I drew up my Monday to Sunday timetable, framing the week; my canvas, if you like. I generated it in Excel, spent time formatting the headings, making the day headings bold and providing neat lines for the table grid, essentially titivating the position of my canvas on the easel.

I hesitated when it came to typing anything into it. I looked at the training Kelly had done previously before looking back at my blank canvas. I repeated this process over and over. I put some of my ideas down, changed them, upped the number of reps and the distances of each rep, then changed them again and scrapped some ideas, continually adapting them.

This was my version of writer's block. The reality was that I wasn't ready. I didn't have enough information to act with certainty. I needed time to generate ideas and I needed to see her training in more detail.

I went back to Aston and Kelly and said: "I'll write the running programme. However, if I'm going to write it, you'll be looking to me to account for the results. Therefore, if I'm taking responsibility for your results, I've got to know what I'm working with."

I explained that I wanted to spend at least a two-week period reviewing all her training. I needed her to allow me to map what she did and how she did it, together with her responses.

Aston and Kelly explained that they were going to a training camp in Monte Gordo, Portugal, for the next few weeks, so that would be tricky. They invited me to accompany them, but assumed I was probably busy at such short notice.

"Portugal, you say? Next few weeks? That will be difficult. I'll see what I can do."

What followed was one of the most brutal calendar culls of all time. Before I could say, "Faro airport," I found myself packing bags for a two-week training camp in Portugal.

RAISED EYEBROWS

Once we had settled into the camp, the measurements began. Kelly made herself available for monitoring every morning and evening. For each session, I applied a novel (at the time) GPS system and measured heart rate, lactate, blood chemistries, sweat rates and blood haematology; the works.

Kelly was her exuberant self at the camp. She was on the front foot with some cutting banter about why we were doing what we were doing when talking to me, but staunchly defended why we were doing what we were doing when talking to anyone else. She questioned everything. She not only needed an explanation of the measurements we were taking, but also of how they would help. I could see her processing explanations and weighing them up in her mind.

I gave as much information as I was prepared to, but I knew I was recording more information than I normally would to give me the complete picture. Of all the athletes I have worked with, Kelly is probably the best-versed in explaining a broad range of applied scientific concepts. She absorbed all the discussions like a sponge. She would digest the information, put it into her own terms and regurgitate with sound scientific accuracy. This is a superb position to be in, because not only have your ideas landed, but the athlete has taken ownership of them.

Indeed, many of the coaches and athletes at the training camp raised eyebrows about just how much scientific activity was going on around Kelly. If she was questioning what I was doing, then many of the other coaches were also asking questions; in their heads, amongst themselves, and directly with me over breakfast, lunch, dinner, at the training ground and in the gym.

I sat down with Kelly and Aston at the end of the training camp and took them through what I had found. It was clear to me that the training sessions Kelly was performing were written to do the right thing, but that the way she delivered them was not. For example, Kelly undertook interval sessions designed to address 800m fitness, but she was performing them too slowly. The same was true for the 200m oriented session. It was structured in a way that caused her to fatigue greatly. The training was intense and she would be fatigued at the end of it, but it wasn't hitting the spot.

The speed that she needed to perform at for an 800m was 22 km/hr for the two-minute event. For the 200m it was more like 28 or 29 km/hr. Kelly was performing her 200m training sessions at an average speed of around 24 km/hr, and her 800m training session at a mean speed of 17 to 18 km/hr. This just didn't make sense. It wasn't specific enough. The training sessions looked good on paper, but they weren't structured or delivered in a way that was specific to her event demands. With limited training time that can be dedicated to the running events, specificity is critical, especially at the business end of the season.

I presented the case for Kelly to take on a different way of training for these two events. I also made the case that, while her top-end speed was good, she wasn't pacing it properly. Part of my argument created the case for her to be able to spend more time developing her sprinting. The biggest challenge I could see was that she lacked fundamental fitness, and with only four months to go before the World Championships and sixteen months to go before the Beijing Olympics, Kelly would need to go backwards to go forward. Endurance training was a necessity, otherwise there would be a limit to the amount of progress she could make within that season.

This focus on endurance was the biggest obstacle for me in convincing Kelly, Aston and the rest of Kelly's coaches. I even got a call from an old boss asking me what the hell I was doing working with a heptathlete! So I began to draw up a battle plan with three points:

1. Craft each session and then discuss all components with Kelly.

2. Attend all sessions, observe everything and capture appropriate measures of her physiological responses.

3. Collate all the relevant information and arrange regular updates with Kelly and Aston.

Extremely conscious of the potential for the opposing stimuli to pull physiological abilities in different directions, I began to create the argument for Kelly to enter an endurance block in the pre-competition phase. I recognised I was skating on thin ice.

As we had numerous experienced coaches on board, they would all feel the same accountability for their work and Kelly's results. Therefore, given that the events in heptathlon are predominantly high force or impulse- and speed-based, such as high jump, long jump, shot-put and javelin, the overall argument for endurance training appeared weak. I not only had to convince Kelly and Aston, but I also had to convince the other coaches. I knew this wouldn't be easy, and I was fully expecting them to react, respond and oppose my idea. And so they did, mainly on the basis that I wanted to "turn her into a marathoner"!

Kelly, however, understood. She had grasped the meaning of her physiological profile showing a mediocre aerobic capacity and had noticed how prone to fatigue she had been in her running training and performances during the big early season event in Götzis, Austria. Kelly was convinced. She supported my argument and knew this had to be done if she was to get the most out of her physiology. Several of the coaches made it clear that they would be watching her every move to see whether there was any noticeable decline in her performances in their events.

Figure 21. Kelly Sotherton undertaking cycle training while she was carrying a small injury in Formia, Italy, in 2009, with me looking on and Tom Parsons (the high jumper) getting ready for a training session.

This questioning and scrutiny was entirely understandable and prompted me to question my own judgements, and further changes often resulted from this. On the one hand, you could say they were taking a very singular view of their events, but in reality the challenge was helpful to my own work. They were asking questions with Kelly's best interests at heart, and the suspicions arose from me being the new kid on the team, and from having put forward a proposal that ran counter to well-established determinants of heptathlon performance, dynamic jumping and high-speed running. If I had been in their shoes I would had raised the same questions.

Applied science can get you into these strange situations sometimes. The steps of reasoning can take you down a path such that, when you arrive, you're not entirely sure how you got there: taking some knowledge, adding some research, building it into an idea, listening to athletes, listening to coaches, weighing up all the factors and then taking the all-important leap of faith. Just as you're about to step off the ledge, you need to check, how did I get here? Am I sure this is the best way to proceed?

I was certain, but I still proceeded tentatively. I asked myself, if you were to design a silly scenario in which a physiologist who was used to working with middle-distance endurance athletes started working with a 'powerful' athlete, what would be the silliest thing to focus more attention on? The answer was probably endurance.

With this continual questioning and uncertainty, I added a further, very important, principle to my work. I needed additional measures to be recorded by Kelly's strength and conditioning coach, and by her biomechanist, so that I had further observations as to whether my sessions were having the desired effect or whether there was evidence that my sessions were detracting from high-force production/impulse development. I needed to provide myself with an assurance that I was not only doing a good job for Kelly's running, but that I was also committing to the overall team goal. I needed to make sure I wasn't dragging Kelly down in other areas, as the classic see-saw of strength versus endurance couldn't be underestimated. I had to make sure that the ingredients I was adding to the recipe weren't spoiling the overall dish.

So, the endurance training began, the anaerobic training began and the sprint training began. The first objective was to segregate the running training into three distinct areas of aerobic, anaerobic and all-out sprint training. This was something we could achieve immediately. The second objective was to steadily increase the running training so that Kelly simply became fitter by doing more; not a huge amount more, but enough to develop her fitness. The third objective was for Kelly to execute her training with greater precision, and this took her a bit of time to get used to.

Kelly tended to go as hard as she could for all her training sessions, which was admirable at the very least. However, I had discovered with the GPS and blood chemistry analysis that this pacing method simply muddled the intensity at which she delivered her training. It also had the consequence of ensuring that her running training was high-effort but not specific.

My advice was based on the profile of her speed during running repetitions, where she went off very fast and then fatigued, with her speed waning during the second half of her runs. This meant her mean speed wasn't high enough and she wasn't pacing the effort in line with what she needed to deliver for optimal 200m or 800m performance.

Kelly needed much more convincing around this third objective. She felt as though she was short-changing herself by not 'going hard' from the beginning. However, Kelly was shrewd and quick-footed, and she could see that her mean speed was much better when she paced her running well. There is nothing quite as persuasive as quick times to reinforce a concept! Specifically, she needed to go hard at the start of the repetition but, crucially, only hard enough to get up to the desired speed, or actually just slightly above the desired speed. Then she needed to hold the desired speed steadily for as long possible. Even pacing gives far better results than volatile pacing.

Kocs

I was recording Kelly's physiological responses for each session and they were guiding some of my judgements. However, if I were to evaluate the extent to which I was using physiological data to inform and direct versus the cumulative observations of the stopwatch, running technique and athlete feedback, I would say it was a ratio of ten to ninety for physiology versus coaching information, respectively.

"What? Are you coaching Kelly's running now?" I would be asked.

"No, no, no! I'm her physiologist," I would protest in a defensive tone.

"But aren't you already writing her training sessions? Now you're here at the track, supervising her sessions. Isn't that coaching?"

"No, I'm here to ensure that Kelly is doing the sessions properly so she's getting the right physiological effect for the sessions we've designed," I would protest, with my pedantic, nuanced and particular presentation of the facts.

While I was certain this angle on the situation was a fair representation, I knew deep down that I had fundamentally stepped across the boundary for an applied scientist into the role of a coach. If I was writing sessions to address 800m and 200m performance, yet also controlling Kelly's sprint training, then I was coaching. These were not the normal tasks and duties of an applied sports scientist. These were the responsibilities of a coach. In addition, I was responsible for the outcome of the training on a day-to-day basis and, crucially, I was accountable for the competition results. My decisions were informed as much by Kelly's competition schedule as from the information arising from her training performances.

My colleagues, Dan Hunt, Dr Matt Parker, Simon Jones, Peter Keen, Dr Barry Fudge and Dr Jamie Pringle, to name a few, have, to different extents, transitioned from being an athlete's physiologist to being a coach. Dan Hunt and Simon Jones are followers in Peter Keen's footsteps of taking on full coaching duties for British cycling athletes. To say they've been successful is an understatement as they have coached numerous Olympic gold medallists between them. Matt Parker and Jamie Pringle have taken on coaching roles alongside other coaches or have served in specific coaching roles, both to National and Olympic level success.

Barry Fudge provided Mo Farah with extensive physiological support, working with a remote coach, managing and supervising training sessions, managing Mo's day-to-day plans and ensuring the delivery of Mo's competition regime. Barry later moved from being head of science at British Athletics to the head of endurance, a role in which he would oversee the work of various other coaches. Barry has been pivotal to Mo's success in becoming multiple World and Olympic champion.

In different ways, each of these scientists has stepped across the same boundary I did with Kelly in taking on coaching responsibility and accountability. It is an indicator of modern applied science that practitioners are working so closely with the coach that they are often capable of stepping across the breach.

The services that sit closest to the core functions of the coach are most likely to take this step into a coaching role. The physiologist serves to develop the maximum possible energy turnover, so in 'engine' sports such as triathlon, cycling, running, rowing, a physiologist could potentially step into a coaching role. The biomechanist serves to advise movement technique, so in technique-based sports, such as archery, gymnastics and jumping sports, the transition is feasible. The performance analyst advises on the delivery of performance, such as tactics and critical moments, so in tactical sports such as rugby or squash, the transition is also possible. The strength and conditioning coach will develop strength, impulse, aerobic and anaerobic (often labelled metabolic) conditioning, so in sports for which there are strong gym and cross-training elements, the transition is entirely plausible. Physiotherapists are known to supervise training for months on end, but full responsibility for a training programme is less common. Doctors, nutritionists, psychologists, podiatrists, lifestyle practitioners and engineers are professions that don't fly as close to the training programme, so they are also less likely to step across into coaching unless they already have a sideline in coaching.

The word 'coach' originates from the sixteenth century, where, in the Hungarian village of Kocs, wagons and carts were manufactured and sold. Students would travel from Vienna to Budapest by cart with their tutors. The word 'kocs' became synonymous with the cart, and the word evolved from Kocs to coach, as in a coach and horse.

The word surfaced in the English language during the nineteenth century among Oxford University students who used the term 'coach' as a metaphor, and possibly as derogatory slang, for the process of a student being supported and tutored and carried towards their exams. The term stuck and was soon taken up in relation to athletic trainers, no doubt because the metaphor so clearly translates to the athlete being supported, tutored and carried toward a competition. As the stigma subsided in academia, it became increasingly accepted in sport, with the guiding tutor becoming more commonplace.

If the set-up in elite sport were transposed onto the original setting of students travelling by horse-drawn carriage from Vienna to Budapest via Kocs for exams, applied scientists would have helped design a route, ensured that the wheels were well maintained or that the student had the correct food provision for their journey and their exam.

My early attempts at denying my status as a coach were essentially a façade. As my time with Kelly progressed, I ventured fully into working on technique and mental approach, which cemented my role as a coach for all aspects of her running.

Upon receiving the official IAAF start list for the heptathlon at the Beijing Olympics in 2008, I saw my name listed simply as "Coach, Steve Ingham (200/800m)." Whether or not I had accepted it in my own mind or in discussions with others, here was the evidence that I was coaching. Now I was the tutor accompanying the student on that journey, preparing her, schooling her, facing the results and being one of the first faces she would see when the exam results were made known. If the sweat patches under my arms were anything to go by, this was a completely different grade of responsibility from being an applied scientist.

Like the Budapest-bound students, we knew exactly when the test (Olympics) would take place and which other students (competitors) would be taking the test at the same time, but, unlike most exams[xix], we also knew exactly what the test would involve (event demands). For an amateur coach, the process of supporting an athlete might be the sole motivator. There are many people who give up their evenings and weekends, travel extensively and invest their own money into aspiring athletes, simply

---

[xix] Unless you get hold of the manuscript beforehand! See chapter two. Naughty!

for the love of it. Equally, there are many coaches who are outcome-focused and invest their time, effort and commitment for the athlete *and* for their own success. Indeed, such coaches make the world of sport go around.

However, coaching can be a lonely place. You feel the weight of responsibility on your shoulders. I know several coaches who have become fascinated to the point of obsession by the process and the outcome. It can become a compulsive addiction to think hard, iterate continually, search incessantly and intensify the experience. I have known many coaches who have an unhealthy balance to their lives, resorting to excessive alcohol intake to numb the stress and quieten the mind as the pressure mounts and the burden becomes too great.

For those charged with the responsibility of almost every decision at each step of an athlete's journey, the time leading up to an Olympic final is all-consuming. The weeks and even months before the Olympic ultimatum are equally capable of creating strange, unusual and unpredictable behaviours.

Alarmingly, coaches have turned to me two days before an Olympic final and asked me if they should change the warm-up! This is the worst time to meddle and almost certainly a symptom of straightforward desperation. I have also had coaches fly off the handle when data has shown their athlete to take a downturn, which has implications about the effectiveness of their training programme. Unsurprisingly, showing your calibrations in such a situation does little to calm a coach's temper! Such aggression is most likely brought on by the need to shift responsibility, offload some pressure and act as an outlet for their competition-induced state of mind.

| 1837 | **SOTHERTON Kelly**   GBR | 31y | 273d | 1976 | - | **6547 -05** |

2004 Olympic Heptathlon bronze ... 2006 World indoor silver ... 2006 Commonwealth Heptathlon gold

pbs in Heptathlon events:

100H: 13.21; HJ: 1.88/1.87; SP: 14.66; 200: 23.40; LJ: 6.79; JT: 40.81; 800: 2:07.94

2006 ECP SP 2003; 3 OLY 2004; 2 EIC 2005/2007; 5 WCH 2005 (8 LJ); 1 COM 2006 (2002-7); 7 ECH 2006 (dnq LJ); 2 ECP 2007; 3 WCH 2007; 2 WIC 2008. As a teenager she represented Hampshire at Netball ... Isle of Wight, lives-Birmingham ... owns four pet cats ... her coaches: Aston Moore (Co-ordinator & LJ), Arun Singh (strength), Steve Ingham (200m & 800m), Graham Knight (hurdles), Fayyez Ahmed (HJ), Shaun Pickering (SP), Mike McNeill (JT) ... 1.78/66kg

In 2008: 1 UK Indoor LJ (3.74, 6 SP); 2 WIC Heptathlon; 4 Long Beach SP; 11 Mt SAC Relays SP; 2 Leiden; 1 Leiden 100H; 5 ECP 4x400 (51.01 split);

3 Biberach 100H (4 200); 3 UK Champs HJ (2 LJ, 9 SP, 11 JT); 1 London Grand Prix 4-Event Challenge

| 1872 | **KESSELSCHLÄGER Sonja**   GER | 30y | 206d | 1978 | **6311** | **6311 -08** |

Sixth at 2004 Olympics ... pbs in Heptathlon events:

100H: 13.34; HJ: 1.85; SP: 14.94/14.53; 200: 24.52; LJ: 6.42; JT: 46.06; 800: 2:11.92

7 WJC 96;3 EJC 97; 3 under-23 ECH 99; 2 ECP 2001 (2002-3); 3 WSG 2001; 5 EIC 2002/2005 (2000-6); 9 ECH 2002; 4 WIC 2003/2007; 8 WCH 2003 (2005

-10, 2007-13); 6 OLY 2004 ... coach-Klaus Baarck ... 1.78/65kg

In 2008: 9 German indoor LJ; 14 Götzis; 1 Neubrandenburg LJ (2 SP/LJ); 3 Ratingen; 8 German LJ

| 1899 | **SCHWARZKOPF Lilli**   GER | 24y | 351d | 1983 | **6536** | **6536 -08** |

2006 European bronze, secured by winning the javelin with 51.36 ... fifth at 2007 World Championships

pbs in Heptathlon events:

100H: 13.50; HJ: 1.83; SP: 14.83/14.26; 200: 24.78; LJ: 6.35/6.34; JT: 54.81; 800: 2:09.63

5 WJC 2002; 2 under-23 ECH 2005; 3 ECH 2006; 5 WCH 2007 (2005-13); 1 German junior 2004 ... coach/father-Reinhold Schwarzkopf ... born-Kyrgyzstan ...

1.74/65kg

In 2008: 1 German indoor; 7 Götzis; 1 Ratingen (including pb 54 81 Javelin)

Figure 22. 2008 Beijing Olympics IAAF official start list. Athlete 1837. Listed were her many coaches, including myself.

BEIJING JUSTICE

It was a heart-warming yet relieving moment when I got a call from Kelly after her second 200m race under my coaching.

"Yeah, all right. I suppose your training works," Kelly conceded when the news of a personal best came in.

Kelly's running results continued to improve throughout the 2007 and 2008 season. In the 2007 season, she set four personal bests in running events, including two at the Osaka World Championships. Throughout the 2008 season, Kelly set eleven personal bests in running events, including the hurdles, 200m and 800m at the Beijing Olympic Games. Her running improved so much throughout the 2008 season that she was drawn upon for the 4x400m relay team, and ran in the heat and final of the event at the Olympics.

On the one hand, you might say my stint as a coach was successful. Fifteen personal bests across a full range of running events isn't bad. Kelly would say that she wouldn't have been as good an athlete had she been left to her own devices, without the science-based coaching approach. I certainly didn't shirk my responsibility or accountability for her results.

The rigorous approach I took in ensuring sound measurements, observations and a rational, objective approach most definitely gave me confidence. Simply put, I knew where she stood at every step and every moment. Without GPS data, heart-rate responses, blood chemistry responses, and, of course, Kelly's feedback, I would not have been able to predict her 'form' and therefore foresee her breaking new ground. It certainly would have been a whole lot more unpredictable and unnerving.

Accepting praise for the work I carried out to improve Kelly's 200m and 800m performance might be reasonable. You could even say the work improved her overall heptathlon performance, given her new best in 2008. The personal bests achieved throughout her events, in addition to the data showing the overall balance between strength, power, anaerobic and aerobic emphasis in her training was appropriately balanced.

Kelly should have won a medal on the day in 2008. Her high jump was below expectations, and since it's the second event in the competition, she found herself playing catch-up and carried the disappointment into her shot-put. By the time the long jump came around the next day, Kelly was too tense to perform to her maximum. Ultimately, she just missed out on the day. As a team, we did not deliver Kelly Sotherton to that competition on that day, and Kelly did not execute to the best of her ability. So, in accepting the role as coach and the responsibility and accountability of the result, I, along with the whole team, should ask, 'How could I have done better by the mission to deliver a medal?' That's what accountability is all about. If you are taking your work seriously, you have to answer to the results.

On the day in 2008, Kelly crossed the finish lines, in both the heptathlon and the 4x400m, in fifth place. Just a few days later, her heptathlon fifth was upgraded to fourth as a previously doped athlete, Lyudmyla Blonska, doped again, was caught and lost her silver medal. The same occurred in the 4x400m. On the day, Kelly, along with Marilyn Okoro, Christine Ohuruogu and Nicola Sanders, finished fifth. However, a series of retrospective tests showed that teams finishing ahead of them had used doping agents to enhance their performance. Russian athlete Anastasiya Kapachinskaya's samples were found to contain banned anabolic steroids. The Russian relay team was struck from the record books and its members lost their silver medals. The British team was promoted to fourth. So Kelly found herself holding fourth (the most frustrating finishing position) in both Beijing events. But that wasn't the end of the story.

Figure 23. Kelly Sotherton running a lone 800m at the Beijing Olympic Games at a new personal best of 2:07.34.

Eight years on, Kelly and the team were rewarded for their efforts with a bronze medal in the 4x400m relay event. In November 2016, Belarusian athlete, Sviatlana Usovich, tested positive for the banned substance dehydrochlormethyltestosterone (turinabol for short). Consequently, the Belarus team was disqualified, lifting the Brits into a medal position. Then, in what was becoming a ridiculously repetitious set of events, Russian heptathlete, Tatyana Chernova (remember her from the last chapter?), had her bronze medal stripped from her nine years later after further reanalysis of her urine sample also showed traces of turinabol (dehydrochlormethyltestosterone for long). This meant that, within the space of a few months, Kelly became a two-time, and then a three-time Olympic medallist[xx].

The 4x400m and heptathlon upgrades provoked obvious delight, but she was understandably upset at not having been justly awarded and recognised on the day.

In response to the 4x400m award, Kelly tweeted: "I've just learned I could be a double Olympic medallist. I'm happy, absolutely f****** fuming and sad."

In the end, sadness and hope are the lasting emotions. It is sad that athletes are resorting to illegal and unethical practices, and that national federations are orchestrating sophisticated programmes to dope while trying to avoid detection. It is also sad that athletes dedicated to clean sport and improving performance through legitimate means, utilising objective measurements and scientific thinking to ensure they develop their own natural gifts, have and will be denied suitable recognition for their accomplishments.

But there is hope, in that the authorities are not letting up and are willing to punish retrospectively if needs be. There is also hope that athletes such as Kelly have had the courage to speak out and continue to bang the drum, such that the anti-doping authorities feel the weight of expectations from governing bodies and the athletes themselves. Without this outspokenness, we would almost certainly still be in the dark ages.

LOYALTIES

As it was, we didn't have an almighty clash on our hands between Jess and Kelly at the Beijing Games. Not only did Kelly miss out on the podium on the day, but Jess didn't

[xx] *All subject to potential Court of Arbitration for Sport appeals, of course.*

even make it onto the plane. Jess failed to complete the big, early-season, multi-events competition in Götzis. She had to withdraw owing to an ankle injury that was probably sustained in the weeks leading up to the competition but flared up in the hurdles and high jump competition, thus ending her Olympic hopes that year.

Rewind to twelve months earlier and I was in a position of truly split loyalties. I had been on Team Jennis since 2005, so I had worked with Jess for two years and seen her progress at a rate of knots. In 2007, Jess stepped up from the 2006 Commonwealth Games to the 2007 World Championships. She just missed out on a medal, beaten by Kelly Sotherton. So I had more history with Team Jennis. I had acted as a support scientist, advising, suggesting and recommending my work to Jess and Toni. I had spent less time with Kelly – just a matter of months – but I had gone beyond the applied scientist role to coaching her performances in running up to her World medal in Osaka.

As I followed the 2007 Worlds on television, waking up extra early to see the events unfold, I was genuinely and completely split in my loyalty. Who did I want to win? Who would I rather see succeed? Would it come down to the depth of the relationship? Would it come down to the length of the relationship? Would it come down to which athletes and coaches I preferred working with?

It goes without saying that I wanted to see both athletes step onto the podium. It goes without saying that I wanted them both to beat Carolina Klüft and Lyudmyla Blonska (the aforementioned doper). However, I really didn't want either to succeed above the other. For each event, I knew which performances they should expect and I was constantly calculating whether one had the upper hand over the other. Ultimately, I had reconciled this in my head before the event started. I would fully accept whether one won over the other. It was easier to divorce myself from them personally and deal with the outcome in a cold, objective way. It was far, far simpler when I became clinical about it.

This way, I could navigate my way through the divided attention and loyalty by simply wanting each athlete to realise her own personal best for each of her events. I wanted to see each stretch to a new level in keeping with the results her hard work had warranted and was right for her stage of development. As a senior athlete, Kelly was trying to move up by a small margin, while Jess was a rising talent who was rocketing onto the world scene.

The issue wasn't just focused on me and where my loyalty lay; it manifested in how I was perceived by each camp. I had been drawn in closely to work with Jess and Kelly, to different depths and in different ways. There is no doubt that they both perceived this as a threat to their own progress. Kelly latched on to the support and development I was giving Jess and had seized the opportunity for her own gain.

Equally, Jess and Toni continued to utilise me during the 'Kelly years', but in reality they didn't need me as much as Kelly did for her running. Toni was overseeing all aspects of Jess' training and my role was simply to advise, support and recommend ways in which they could progress.

While I was working with Kelly, Toni certainly changed the way he worked with me. The relationship became more guarded when it came to revealing and discussing his ideas and preparation plans. If I had been him, I would have done the same thing. I would simply have kept me at arm's length from the granular details and used me for consultations and discussion. Anything more than this and I expect he would have imagined I would be taking the ideas and implementing them with a rival.

In fact, this couldn't have been further from the truth because Jess and Kelly were distinctly different. Yes, they are both runner-jumper types; however, their underlying individual physiology, which determines how and why they are good at what they do, was different. Of course, there are some generic areas of preparation that all athletes can follow, such as a sensible training load, use of resistance training and a sound recovery plan, which would apply to both athletes, but this advice would apply to all athletes.

As you are dealt a different set of cards with each athlete, you must adjust the way you play them based on the variables in front of you. With both the generic and the individualised advice and support, the race between two rivals is not dependent on whether one knows and copies what the other is doing. The advantage comes from how well the coach and athlete know the athlete, and how well they integrate and utilise the advice available to them. Alongside everything a coach and athlete do, they also are accountable and responsible for the application of science. Scientists should also bear the responsibility of athlete results, whether as part of the multidisciplinary team or in stepping up to a coaching role.

Kelly pronounced her needs by challenging me to have a positive impact on her goals. The partnership Kelly and I formed taught me:

- That my fundamental training had been enhanced by my own experiences as an athlete, but my empathy and understanding for the role of the coach was infinitely deeper having been one at a high level.

- That assuming responsibility for Kelly's results had added the full weight of accountability to my work, leading to increased ownership and therefore preparation.

- That I needed to develop a continual evidence base and receive high-quality athlete feedback about progress in order to make confident, assured performance decisions.

Chapter 6: Beware of your Thoughts…

"Leadership is not about a title or a designation. It's about impact, influence and inspiration. Impact involves getting results, influence is about spreading the passion you have for your work, and you have to inspire team-mates and customers."

Robin S. Sharma

Two Pair

We've been travelling through time chronologically so far in this book, so that you can hopefully see themes and lessons building that have better equipped me to take on the next challenge in my career.

Now I need to go back in time to an earlier encounter because I want to share with you a moment in my support journey that floored me. That moment is most relevant to the narrative of this part of the book.

Let's go back to the 2000 Sydney Olympics to warm us up. The year 2000 was a landmark in time and in British sport, heralding a massive step forward in performance standards. In the background, the sporting institute systems were forming, lottery funding was coming on stream and the Olympic rowing team had been reborn from the depths of thirty-sixth place at the previous Olympics in Atlanta and was once again the pride of the nation.

Before I flew home from the Games – after the highs of seeing the rowing team celebrate three medals – there was the small affair of watching the men's and women's 100m final in Sydney's Olympic stadium and the following night watching the most incredible sequence of athletics that has possibly ever occurred.

On 'Magic Monday' – September 25th 2000 – I witnessed Jonathan Edwards win his gold medal in the triple jump, Anier García win the 110m hurdles final, Sonia O'Sullivan and Gabriela Szabo go head-to-head in the 5,000m final, Haile Gebrselassie and Paul Tergat tussle all the way to the finish line in the 10,000m final, and Michael Johnson crown a glorious Olympic career with another win in the men's 400m final. Then I joyously watched as Cathy Freeman sent 112,500 people into a crescendo wave of celebration and euphoria around the stadium in Australia, along with countless others across the globe. Her performance transcended sport and gave society a new dimension to its understanding of equality, heroes and responding to adversity. It was a true privilege to watch and soak in.

A few months later, we were back to business as usual. My duties continued. I was still supporting the rowing team, focusing on the men. Macca was still around, picking up the pieces of the men's 8+, which had been riven by retirements. Jürgen was as committed as ever, but was experiencing a very slight decline in intensity during the post-Olympic year. We had a new recruit at the Olympic Association in the form of Rob Shave, who had been brought in to help support the women's programme. The testing, the trips to the river, the meetings with coaches and the training camps continued as normal.

Figure 24. Matthew Pinsent training at altitude (2,300m) in Sierra Nevada, Spain, while I took a blood sample.

As many were experiencing Olympic lulls in terms of their training, some were still partying and enjoying their Olympic medals. James Cracknell, on the other hand, barely broke step. He was straight back into training, preparing himself for the following year and heading towards the Olympics in Athens long before most people had even realised where the next Games were to be held.

I would generally describe Matt Pinsent's physiology as phenomenal, and I wouldn't be the only person to do so. James Cracknell was seen to be shouting, "You horse!" to Matt after the 2003 World final in exultation of Matt's incredible horsepower.

James wasn't far off Matt's standards, but he didn't have quite the same hardware. However, James took the approach that if he didn't have the extreme hardware of a Pinsent, he would do the best he could to programme his own. James had an insatiable hunger, an extraordinary diligence and a drive to look after himself, which meant he had a healthier immune system to recover more efficiently.

He was, and still is, the consummate professional, which forced his system to become outstanding. For much of his career between 1997 to 2004 he was at the limit of what his body was capable of. The fire of dedication would persist long after his Olympic flame expired. Duathlons, marathons, rowing the Atlantic Ocean, the Yukon Arctic Ultra, Land's End to John O'Groats; you name a bonkers ultra-endurance event and James has done it or is probably planning it. He is one focused chap. I remember Steve Redgrave telling James he was being a bit intense. If Redgrave tells you this, you know you're off-the-scale focused! But it was this focus that enabled him to compete and match the incredible horsepower of Matt Pinsent.

The year swung round to the summer of 2001 and the major news was that James Cracknell and Matt Pinsent were teaming up in the coxless pairs. They began their season with great success, winning all three World Cup events entered in New Jersey, Seville and Vienna. They were head and shoulders above the competition, and they knew it.

That season, Matt and James were hungry for a new challenge. They had switched to the pair but realised the competition, and therefore the challenge, perhaps wasn't as great as they needed it to be. They began to hatch a plan to up the ante. It wasn't until the middle of July, when I met with James and Matt, that I heard about the new challenge. They just dropped it into the conversation as casually as if they were planning to pop down to the newsagents! Their idea was to enter both the coxless pairs and the coxed pairs events.

Traditionally (it's different now, with the finals spread more evenly throughout the week), all the rowing finals took place on the final Saturday and Sunday of a regatta. Performing in two finals on two consecutive days would be hard enough, not least because the preceding days would see them undertaking two sets of heats and semi-finals. This meant they would have double the amount of fatigue going into the final weekend before they raced the finals. If this wasn't enough, the finals of the coxless and coxed pairs were timed to begin two hours apart!

The nonchalant approach to them considering it was thrown at me by James: "We think we might need a recovery plan in between the finals. What do you think?"

"Err, YEAH!" was my initial response!

Then I added: "Well, is this something you're going to do or not? Because if it is, and it's something you want me to work on, I'm going to have to do some serious packing so I have all the necessary equipment to be able to answer your questions!"

Off we went up to altitude once more, to the training camp they all find so tough, but that prepares them so well for peaking at the World Championships or Olympics. This time I had a new challenge, and at this early stage in my career it was a fantastic opportunity to get my teeth stuck into a really interesting physiology project with a clear performance outcome.

WINDOW FRAMING

I sat down with Jürgen, James and Matt early in the camp at Silvretta to take them through my thoughts. I would supplement the normal baseline morning monitoring with additional training measurements. I proposed that we should add further measurements, topping and tailing their sessions to put their pre-session and post-session state under the microscope.

Would it be better for them to row to recover between finals or would it be better for them to cycle, run and receive massage, or simply not do anything between finals? They would need to stock up on fuel before the first final to deliver not only the fuel necessary for the initial event, but also to set themselves up for the second final. What could they eat or drink between finals to replenish themselves? What food would they find palatable before and immediately after a race? The event was held in Lucerne, Switzerland, and while Switzerland isn't wildly hot, the microclimate around Lake Rotsee was notorious for being humid. So how would they cope with the heat?

As we went through the full scenario, allowing for unforeseen circumstances, we estimated that we only had about an hour to intervene because, for the first seven minutes of the two hours, they would be racing in the coxed pairs event. For the next ten minutes, they would be turning the boat around from the finish line and paddling back towards the boathouse, so they would have limited opportunity to do anything constructive in this time other than row. Before they could get out of the boat, the two hours had become one hour and forty-three minutes at best.

Figure 25. Matthew Pinsent and James Cracknell in the coxless pairs boat at altitude in Silvretta, Austria, after coming in for their physiological measurements.

At the other end of the two hours they would have to wait on the start line, perhaps for five minutes. They also had to row from the boathouse with their coxless pairs boat and position themselves with the starting stake boats.

Before that, they would need to do some sort of warm-up, assuming that in the two-hour window they would have lost some of the priming effects of the previous warm-up and the previous final.

Jürgen was very clear they would need at least twenty minutes to prepare mentally for the subsequent final. All of this would take approximately forty minutes before the second final. Therefore, we had a window of an hour beginning twenty minutes into the two-hour window and finishing forty minutes from the end of the window for us to intervene.

Part of me thought, 'It can't be done.' However, the applied scientist in me knew they were extreme specimens with a clear advantage over their rivals, which, meant they stood a chance. They also wanted to have a go at it. It was their goal, their challenge for the season. My job was to give them the best information about how best to recover, and I would also possibly have an input into deciding whether or not it could be done.

There was only one way to deal with all the issues at hand: to trial and test their responses to different interventions and then try to make the best judgements possible in the six weeks from the start of the camp to the beginning of the World Championships. And that's what we did. After every training session, we plotted their recovery. If it was a low-intensity training session, I would track the recovery of their hydration status and their body temperature. If it was a high-intensity session, I would also ask them to trial a recovery method. For example, we compared just rowing versus cycling, versus walking, versus doing nothing, versus massage, versus a combination of the above. I worked with our team back home, communicating with them via dial-up, fax and Austrian schilling payphones; smoke signals by today's standards.

I worked with nutritionist Jacqueline Boorman to estimate the energy, fluid and electrolytes that needed to be replaced versus how much they could ingest and absorb in the available window. I worked with my boss at the time, Dr Richard Godfrey, who provided me with relevant papers and opinions from other experts, such as the late Professor Tom Reilly, about the rate of heat dissipation and the recovery of metabolism after all-out racing. I was fielding the questions, working with colleagues and acting as a funnel for their information. I felt gratitude then, and have always been indebted since, for a fantastic network of co-workers, colleagues and friends that I can lean on for advice and support.

Camp fever

I spent a considerable amount of time with James and Matt at those training camps, probably four to five hours each day. They would do their normal morning monitoring and stay behind for extra pre-training sessions. We would take extra training measurements and always did post-training session measurements. This meant that we started and finished later than everyone else. We went to breakfast, lunch and dinner together. We would sit down and debrief the results, either after training sessions or at the end of the day, and we also spent time simply socialising together.

Anyone who has ever been to a training camp of more than a week with athletes will know that camp fever inevitably sets in after the novelty of being in a different, perhaps warmer, perhaps more scenic, environment either wears off or gets overtaken by other pressures. The routine and focused attention on training are two of the best reasons why training camps work so well for dedicated blocks of work. However, the by-product is that no matter how much you enjoy training or other people's company, claustrophobia can close in around you.

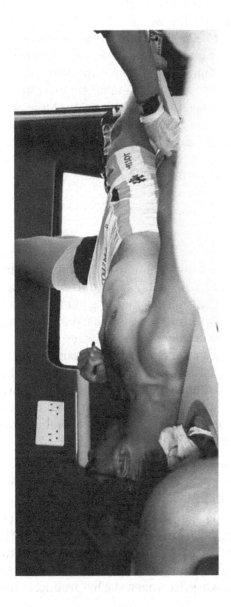

Figure 26. Matthew Pinsent in the back of the old BOMC Renault Trafic van undergoing a bioelectrical impedance test to assess fluid loss.

If you misjudge your packing for one camp, you're likely to go a bit crazy and forever know that you need to come prepared in future. People pack walking shoes, books, textbooks, bikes, telescopes, paints, TVs with satellite dishes, bagpipes; you name it, it has been brought on a training camp to alleviate the frustrated mind.

At the altitude training camp in 2001, it was the project of building a remote-controlled aeroplane that absorbed our attention. Coach John West bought it, and the main construction team comprised myself, Pete Gardner, Rick Dunn, Neil Chugani, Matt Pinsent and various others. We took it from kit to maiden flight in four weeks. The fact that its lifespan as a functioning aircraft lasted only about thirty seconds before the pilot tanked it into the ground is incidental. We bonded as a group as we lovingly built that plane, solved problems together and revelled in the excitement of its first take-off. We were also grief-stricken/had many belly laughs as we picked over what had gone wrong while staring at the smithereens splattered over the concrete in a car park in France. We pretty much played like children, but this play was essential in maintaining team morale, or at the very least keeping camp fever at bay.

QUESTION TIME

During the first three years of my time with the rowing team, I knew I had to work *with* them to build rapport, acceptance and trust, and to be effective. I also knew I would have many questions about why they did what they did. I had resolved to address some of the questions through my PhD studies, which in themselves developed an evidence base for the distance approach so heavily questioned by scientists and critics.

As my confidence and acceptance with the team grew, I began to ask a few fundamental questions that I had accumulated. Over lunch with James and Matt following our post-training testing, I ventured a slightly fruity one.

"Matt?"

"Yes?"

"Can I ask you a question about how you row?"

"Yes," he replied, one eyebrow slowly lifting.

"Well, I wonder why your stroke length is pretty much in the middle of the pack compared with the rest of the heavyweight team, yet you're one of the tallest and have one of the longest arm spans? Shouldn't it be one of the longest?"

I knew I was prodding him with this question. I knew it was bordering on slightly cheeky. And I knew I had hit a little bit of the target when James chirped up: "Ha, he's got you there!"

Matt had a well-polished answer: "When you're rowing a boat and when you're competing, the most important thing is being able to maintain your technique and stroke length, rather than how long it is."

This was a reasonable answer, but it wasn't totally convincing. I probed a little further.

"Have you ever thought about increasing your stroke length to take advantage of your lever length?"

Again, his response was well-crafted. He started slowly, which might have indicated some uncertainty, but he made up for it by giving his answer with an air of teaching: "That's a fair question, but you have to understand that at thirty-one years of age I have to weigh up the risk of modifying my technique to get more out of it, not knowing whether it will be mastered in time to be delivered in competition, *or* sticking with a tried and trusted technique; one that I know has delivered."

I conceded the point with affirmative nods. "Yes, I can see why you would choose that. Thanks. I hope you don't mind me asking. I just see these things in the data and wonder why they are as they are."

"No, that's fine. It's a good question," Matt responded, playing his approval of my question straight down the line.

His response did not give me a green-light invitation to use lunchtimes as an opportunity to quiz them in some sort of cathartic science cross-examination. I resolved to choose my questions carefully. I had another waiting in the wings and chose to roll it out a few days later. This one had been at the back of my mind since the month before I had started working with them.

At the Munich World Cup regatta in May 1998, the men's four had being forced to change their crew as Tim Foster had sustained an injury. Luka Grubor, an excellent rower with a big engine, who would later go on to win a gold medal in the men's 8+ under Macca, had been substituted for Tim at the last minute. Despite his strength, the crew came fourth. The press reported this widely because it ended Redgrave and Pinsent's seven-year unbeaten run.

The men's four had further dominated up to the Sydney Games, yet at the last World Cup in Lucerne before the 2000 Olympics they had lost again, coming fourth once more. On the surface, you could see why they hadn't won in 1998, with a new crew member stepping in, but in Lucerne the slump was difficult to explain. Jürgen attributed it to overtraining, while Tim commented on poor technique and James blamed a short stroke length. However, I was curious to know why they had fallen from a position of consistently dominating and winning all the way to fourth.

I remember distinctly where I was when I posed the question to Matt and James. I was sitting with them at a table on the right-hand side of the Restaurant Silvrettasee veranda overlooking the majestic lake. We had just finished our lunch, which had almost certainly been topped off with a Germknödel, the renowned belly-filling dumpling. I waited until they were halfway through lunch before I got my question underway.

"Can I ask you another question?"

"Yeeeees." Matt's tone sensed there was something meddlesome ahead.

James just shrugged.

"Well, I know how strong you guys are. I know your physiology. I know you're both off the scale in your own way. Matt, your lungs are impossibly large. James, you have a ferocious appetite for training and an amazing ability to recover. It's obvious you're so much stronger than the opposition. You've basically won together since 1997, and Matt, you've won since 1991... Except for in Munich and Lucerne those times. So my question is, why didn't you come second at those World Cups? Why didn't you come third even? If you dominate your opposition so comprehensively and regularly, why is it that on an off-day we don't see a slight decline in performance, but a catastrophic decline in performance?"

James looked over at Matt, silently suggesting that he wouldn't be taking up the answering role. Matt finished his mouthful.

"Well, in 1998, Luka obviously subbed into the boat at the last minute and it really disrupted the rhythm, and in

2000 we were overtrained, our length wasn't right and we were a bit overweight."

Matt trotted his explanation out as if he were drafting his autobiography.

I pushed a little further: "I know there were extenuating circumstances, but do they really add up to you going from first to fourth?"

Matt started his response hesitantly, clearly not ready for this one: "Yes, err, yes, that's what happened. We really struggled in those finals. It didn't come together, we just had a bad day…"

I gave some nods of acknowledgement, a little shrug and a pause, then went again, entirely aware that I was delving a little too deep for their comfort and testing the newfound strength of our relationship.

"To be honest, I'm not convinced. I know your physiology, I know how good you are. I've heard reports about the physiology of other crews, and that matches the performances and the margins by which you beat them. It just doesn't quite add up that you would drop your performance so much that you would end up coming fourth."

"No, no, no, it's not as simple as that. This is an unpredictable sport. It's outdoors. It requires technique. You can have a bad day in the sport," Matt retorted.

I could have responded to each of his points, but I was in knee-deep, so I chanced my arm.

"Could it be that fourth place is a much more convenient place than second or third?"

"What do you mean?" Matt said, recoiling.

"Well, could it be that you didn't want to be on the podium having your picture taken receiving a silver or a bronze medal? That you didn't want the picture in the press or in your mind? Was it just easier for you to back things off and come fourth?"

Matt was already shaking his head, but James gave a pensive, concessionary upturn of his lips, which hinted, just hinted, that I might not have totally missed the target.

"No!" Matt was immediately dismissive. "No, I, err, it wasn't like that. We just had a bad day. It doesn't just come down to physiology numbers you know, Steve!"

This was my time to step off the questioning bus. I sensed a slight tone of, 'Back in your box, you're just a physiologist.' He had said my name and not Hot Buns, Stevie or Stevo, which made it feel as though I was being spoken to by my parents. This all added up to me concluding that this was as far as the conversation was going to go. There was a slight awkwardness; not much, but enough. I broke the tension by quickly asking them if they wanted to look at that day's results, by way of refocusing the relationship to my function and on to the task in hand.

I winced as I got back to my room. I grimaced[xxi] several times that afternoon, contemplating whether I had pushed the question too far. I could not see a way of asking the question, 'Why fourth?' subtly, without me sharing the extension of my thoughts about whether they would effectively opt out, turning their race into a non-event. This left me wondering whether I should have granted the question any oxygen at all. I would ruminate on this exchange for several days to come. I sat down a day later to square off the post-mortem going on in my head

---

[xxi] My favoured face for such an occasion is known as 'the Gary Lineker', a face he famously pulled just as Paul Gascoigne lost all emotional control upon realising he wouldn't be able to compete in the 1990 World Cup final, while mouthing "Have a word."

through some reflective practice (Ailsa would've been proud).

It's one thing thinking about what you could have done better, but as a support provider you're constantly thinking empathetically about what your client, the recipient, the consumer of your work, thinks about you and what you have to offer. I was left to wonder whether Matt or James would be thinking, 'Cheeky bugger,' or even go as far as thinking, 'What does this trumped-up physiologist know? It's totally different for an athlete who has to go out there and perform. He doesn't realise how a performance is delivered. He has no appreciation of how disruptive it is to have a new crew member come into a boat at the last minute and for people to expect you to perform at the same level. And in Lucerne there were a whole host of factors that contributed to us being off-pace.'

COXED AND BOXED

The rest of the camp pressed on. We continued to build the aeroplane, and many more games of Scrabble and table tennis were played. The cows continued to moo and ring their bells outside my bedroom window, but I didn't ask any more of my curious questions.

Within a week, our discussion had drifted from being cringingly front and centre to a yester thought at the back of my mind. The work continued to build towards their double competition and the plan was coming together.

We had a clear idea of what was best for them both, as well as focusing on individualised differences based on their individual responses and preferences. Generally, cycling was a superior option for recovery than walking.

Walking was better than staying in the boat and rowing, and walking was chosen as it was a more practical option.

Eating and drinking immediately was the best option, and we would need to adapt to the competition venue when we got there to see how we could optimise this. Massage didn't influence the recovery physiology but they liked it a lot and, in the absence of evidence that it detracted from performance, subjective approval went a long way to deciding that it would be included. We had built in the flexibility of options depending on what was available at the competition venue and the conditions on the day.

Figure 27. Matthew Pinsent and James Cracknell digesting some feedback about how their physiology responds after exercise.

Once the team had arrived in Lucerne, it was clear that unusually high temperatures were expected. A cunning coach had also spotted an opportunity to get access to the pair through a small break in the trees along the riverbank almost immediately as they were paddling back from the finish line to the boathouse. With five days to go, we made some last-minute tweaks and tried out a new refinement to the plan.

The schedule went like this:

- Time 0:00:00: Race coxed pairs.

- Time 0:07:00: Win coxed pairs (hopefully).

- Time 0:08:00: Turn boat around and paddle easily. Row close to the bank to meet me.

- Time 0:09:00: I would throw two chilled isotonic energy drinks out to James, who sat in the bow seat, while Matt held the boat steady. They would immediately drink the 500ml drinks, then paddle to the boathouse.

- Time 0:15:00: They would arrive back at the boathouse landing stage, where they would be greeted by the physiotherapist (I didn't have access to the landing stage), who would give them a bag each. Of the low glycaemic options presented, Matt chose to have white bread jam sandwiches and some jelly babies, while James would have just jelly babies. They would both have a further 500ml of isotonic drink, this time with additional electrolytes. Other staff would take their boat, so all they needed to worry about was putting trainers on, taking an ice towel and wrapping it around their necks and shoulders.

- Time 0:17:00: Begin fifteen-minute cool-down walk, during which time I would meet up with them to check they had everything they needed. I had spare nourishment in my backpack if they needed it and would accompany them back to the boathouse.

- Time 0:32:00: Finish cool-down walk at the boathouse.

- Time 0:35:00: Each would receive massage treatment from the physiotherapists.

- Time 0:50:00: Bring core temperature down, with Matt opting for further ice towels applied to his torso and James choosing the option of a cold shower.

- Time 1:00:00: Finish the cooling regime, get themselves dried off, get changed into fresh kit and faff about as necessary.

- Time 1:05:00: Relax and refocus their minds.

- Time 1:35:00: Walk down to the landing stage, where their boat would be ready and waiting for them.

- Time 1:40:00: Begin truncated warm-up, focusing on easy, relaxed rowing and forceful priming strokes.

- Time 1:55:00: Line their boat up on the stake boats and relax.

- Time 2:00:00: Begin coxless pairs race.

- Time 2:06:30: Win coxless pairs race (hopefully)!

In terms of a factual record of the event, you could say it was pretty close. However, the emotions and the eventualities of the races were wildly different.

The coxed pairs started uneventfully. James and Matt, with Neil Chugani as their coxswain, led from the start and at every checkpoint along the way. With about 300m to go, they dropped the pace, realising they would win, and eased towards the line. The Italian crew, however, were putting their final sprint together, and Matt and James were playing a game of chicken with how much they were slowing down versus how much the Italians were speeding up towards the finish.

In a heart-stopping moment watching the big screens, I could see that they had only just won, judging their efforts and timing their race to the minimum winning margin of 0.4 seconds. The first third of the plan was ticked off.

The recovery plan began in earnest and went according to the above schedule, apart from me misjudging the first dispatch of drinks, undercooking my first throw (I do hope someone fished it out of the lake!) but pitching the next two to perfection. I joined Matt and James on their recovery walk and then on to the boathouse, where Jürgen had one witty word to summarise their coxed pair performance.

"Bubka," was all he said, in homage to Sergey Bubka, the pole vault maestro, famous for raising the world record by the minimum required each time to ensure he was in regular receipt of the top prize money.

I asked them how they were feeling, provided them with more ice towels, told them to get on with eating when they showed signs that nerves were suppressing their appetite and they said they didn't fancy eating, and generally tried to be a calming force to the frenzy of following the schedule. The second third of the plan had seemingly gone well... tick.

FOURTH

The final third of the plan was by far the zestiest. While James and Matt prepared mentally, got back in their boat and warmed up for the next race, I joined up with Rachel (by then my fiancée) and took my seat next to James and Matt's future wives, Beverley Turner and Dee Koutsoukos, ready for the final final.

Bev and Dee were nervous. I had sat with Dee at the Olympics in 2000, when I had tried my best to calm her down and reassure her that everything was going to be fine; that they were the best-prepared crew, had the best fitness, and would dominate the race. As I did this again in Lucerne, I was holding back my own nerves because I was so much more involved than I had been a year before in Sydney.

Rowing races are difficult to watch for two main reasons. First, there are three main camera angles, which beam footage back to the main screens and the broadcast feed. There is a camera on each of the following boats directly behind the racing crews. There is a van that travels along the towpath with its camera facing the side of the boats. Then, there are fixed cameras at certain points in the race, normally at 500m intervals, which corresponds to the split timing points. The camera angles are rarely perpendicular to the flow of boats, which means you are always trying to adjust for what appears to be an echelon of boats. Unless a crew is considerably ahead or behind, spectators are constantly trying to estimate and judge the perspective and the positioning of the crews. This makes the race difficult to watch due to confusion and uncertainty.

The second feature is the speed and layout of the race. For the first half of the race, you can't physically see the boats. When they come into view they creep slowly down the river. At times, it's a torturous version of 'Poohsticks'. The boats are moving quickly (equivalent to a fast running speed) but, given the distance over which the race takes place, it feels slow until they come into view. Movement between the crews is equally slow, where an advancing crew or fatiguing crew slip forward or backward only stroke by stroke. Both features make watching rowing races extremely tense, especially if you know the people involved.

The second race, the coxless pairs final, did not go according to plan. As the buzzer beeped to start the crews, there was immediately a question mark in our minds. They did not get off to their usual strong start. I thought, 'Perhaps it's just the camera angle. Perhaps they misheard the buzzer or they were held at their starting station.' But the nerves started to rise and the questions from Dee and Bev began. My emotional anxiety hindbrain was hassling for control.

I had worked with this duo for three years, and particularly intensively over the six weeks leading up to this event. I was as informed as anyone in the stands, so I searched my brain for some logic. I reasoned that normal business would be resumed by the 500m mark, so I told myself to be patient and wait for James and Matt to kick through their gears.

The 500m mark came and they were third! They were 0.72s adrift of the Yugoslavian crew in first place. This was an unprecedented margin for them to be behind so early in the race. Now my logical forebrain was struggling to keep up with the body blows coming thick and fast from my limbic system, which was awash with excitement. I was reaching for more ideas and reason. I began to recognise that Matt and James were carrying over more fatigue from the previous final than we had anticipated.

Rather than accept that the double-event ambition was too ambitious, I was beginning to relive all our measurements, discussions, decisions and my recommendations. The recommendations were, of course, based on *my* calculations, estimations and interpretations. I could feel the fingers starting to point. What had I missed? What didn't I know? What was my failure? This escalating anxiety was not helped along as they came into sight, at which point everyone in the crowd leapt to their feet and the wave of cheering began to increase in volume. The screen cleared to receive the 1,000m mark results. They were now fourth, and 1.94s behind the Yugoslavians!

The situation had got worse. I reached into my bag for my notebook. I had an urge to look over again at some of my calculations of thermal loss and to view my estimates of drink electrolyte concentrations. Looking back now, I was unnerved. I wasn't too dissimilar to the person involved in a car accident who, upon collision, begins to adjust the windscreen wiper speed or frantically tunes the radio station in panic.

Rachel looked down at me as I scribbled away with a worried look on her face and asked, "Are you okay?"

I mumbled something about needing to make sure. As much as I tried, mathematics was not going to happen. I was surrounded by cheering crowds, anxious wives and girlfriends, and a vision of the questions looming. Rachel tapped me on the shoulder to alert me to the 1,500m split time that was about to come up. First, second, third flicked onto the screen. Fourth again!

With pencil and notepad still in hand, I attempted to write down the times, but I couldn't read my own writing (nothing new there). I was trying to calculate if James and Matt were closing in. At about 1,600m into the race, I could see the boats clearly. Making my own adjustment, I could see they were still some way down. I would later discover that they were trailing by 1.70s at the 1,500m point.

My mind started to accept the defeat as the most likely outcome. I would cheer as much as I could, but I knew the look of a fatigued racer when I saw one, and I was looking at two. This was added to the knowledge that, when you're already fatigued, the added fatigue accumulates much more quickly than normal. I had resolved to accept that this race was only going to go one way.

A few seconds later, the British crowd began to scream. I craned for a clearer view and could see that the pair had closed from being more than a boat length down to being only a half-length down with 250m to go. As they came past us with 200m to go, Matt and James were only half a metre behind. There was a chance!

We were now fixed on trying to estimate whether they had taken the lead, but as the viewing angle opened up again, we couldn't be sure. The big screen was using a fixed camera at the finish line, so it was no better with the angle offset in a different direction. I was cheering, but it came out as a whimper. We looked up at the big screen with 100m to go, and it looked as though James and Matt had taken the lead. My whimpering grew feebler.

Then, with 50m to go, I could see James 'catch a crab'; rowing parlance for spooning your oar badly into the water during the recovery phase. The boat stalled again, as did our hearts. The Yugoslavians took the lead and there were probably no more than three strokes left. Three almighty attempts to exceed the opposition. The two crews, neck and neck, crossed the line at effectively the same time. This was quickly followed by the crowds gasping their "Oohs" and "Ahhhs", then turning to each other to compare notes as to who they thought had won.

A SEED INTO A REDWOOD

The big screen announced a photo finish for first and second. The results were pending, and with it our sanity and the space-time continuum. Third through to sixth were announced. The minutes passed, sweaty palms were wiped, big breaths were taken and the fidgeting became feverish. The big screen refreshed. The number one appeared on the left-hand side, quickly followed by three letters: GBR! They had won! They had done it! Somehow, they had crossed the line first.

Cheers erupted around us. Hugs, congratulations and big sighs bustled through the relieved crowd. I no longer needed to think through the rates of thermal heat loss for big humans. I no longer needed to be thinking about blood(y) lactate clearance. I no longer needed to be thinking about grams of carbohydrates or the molar masses of potassium and sodium! I was still shaking for an hour later. Not only were my hands rattling, but my head was shaking from side to side, dazed by what I had experienced, pummelled by the intensity of the emotion, and baffled at why I had thought the third quarter of the race was a good time to be recalculating electrolyte concentrations!

We made our way to the hotel to get changed and to meet up with the team in the hotel bar to offer our congratulations. Eventually, James and Matt came through, and there were handshakes, man hugs, thank yous and congratulations. The pleasantries were quickly followed by questions and stories about how the race had felt. James described how he had barely been able to hold himself up to receive a medal at the end of the race. There were plenty of quips from James about the point at which he had used up his last jelly baby.

As everyone continued to mingle, Matt found a quiet moment to put his hand on my shoulder to get my attention. He dropped his voice and then this bombshell, on me: "You know, I thought about you during the race today."

With my eyebrows raised as high as they go and my head reeled as far back onto my neck as possible, I stammered, "What, what, you, what... you did what?!"

"I thought about you during the coxless pairs race," he confirmed.

I was aghast. "What on the earth, were you doing thinking about me during that race? You should have been concentrating on rowing as hard as you could. No wonder you were coming fourth!"

He explained, "I remembered what you asked up in Silvretta, about when we'd lost and came fourth. You were right. I didn't want to get a silver or a bronze, and I don't think the other guys did either. I thought about that conversation when we were trailing in fourth place today! I thought to myself, 'I have a choice. I could choose to come fourth again and the final gets reported as a non-event, or I could choose to fight harder.' So, I decided that if we were going to get a bronze medal or a silver medal, then today we were going to win the hardest-earned bronze or silver we'd ever won. Thank you for asking that question!"

"But, I... what? Eh?" was about as polished a response as I could muster.

I had already had enough to process that day, and now I had a dawning wave of realisation that my words, albeit a bit edgy, had made such an impact on someone, let alone a future knight of the realm! I was torn as to whether I should have asked the question in the first place. I had influenced Matt, but it could so easily have misfired.

With these yin and yang thoughts going through my head, I offered, "Blimey! Wow! Sorry!" to which Matt was obviously magnanimous, gracious and humble in his thanks.

Nevertheless, I was truly struck by how fortuitous I had been. Matt is an intelligent man and I can only assume he had processed the conversation and decided to act positively upon it. Perhaps he had made the promise not to ease off if they were in a losing position weeks before they arrived in Lucerne. Perhaps he had just processed it during the race. My guess is that he at least decided that if he were to find himself in a similar position he would respond assertively to it beforehand, but in the race he had quickly realised that the situation was live, and had grabbed the bull by the horns and acted upon his resolution.

But Matt had done the processing. Matt had done the necessary work to turn this into a positive. Matt had taken the resolve to act, and ultimately Matt had done the pulling on the oar, matched by James, and created the winning propulsion. My brain was being stretched, boiled and split.

Matt's few words of feedback deeply moved me. He had not only heard my idea, but acknowledged it. At the time, I hadn't been overly concerned about whether Matt had processed and ratified my idea. I was more concerned about whether I had annoyed him, having gone a bit too far. I reconciled in my mind that if I had backed off any earlier into the question, it would have aggravated him less but I wouldn't have made the actual point. Now I had heard that the idea was legitimate. It had been recognised and endorsed against all of Matt's experience.

Acknowledgement was only one stage. Matt also had the good grace to inform me that he had taken the seed of thought and let it grow into a sapling. The most I could have imagined was for him to affirm my idea. He could have simply said that my line of thinking wasn't too far from the truth. But among this melee of thoughts, I could not have imagined that it would fully manifest itself beyond the sapling to become a fully-grown Redwood and impact on his performance. The deeply surprising aspect of his admission was that our discussion had been replayed in his mind during the race. It still fills me with humble bewilderment that he was thinking about our discussion and my idea while trying to win.

In 2010, when running an applied science development meeting, I had invited Ben Hunt-Davis, from the Olympic gold medal men's 8+, to give a talk to my team about his Olympic experience. I had briefed him that this was the first Olympic and Paralympic cycle for some of the team. I had also briefed him on the need to talk about the excitement of the Games and how practitioners can be a force for good in that environment by being a steady, calm, likeable person to be around. As a precautionary note, I had delicately asked him not to pepper his talk with too much reference to our work together, principally because I had previously experienced a cringeworthy instance with an athlete, which, while on the surface was pleasant, really seemed like I had gratuitously set it up to aggrandise myself.

Ben certainly did as he was told. He perfectly illustrated the crazy circus of excitement that comes with the Olympics, telling stories of the kaleidoscope of amusements, sport superstars at every turn, the temptations of the dining hall and the general hype in the Olympic bubble.

He then proceeded to tell my team how he couldn't have cared less about what I had been doing during the competition. He said that I had been the last thing on his mind, and that generally he wouldn't waste brain space on support staff when the heat of battle commenced. While it wasn't pleasant to hear, it was what I had expected to happen.

ATHENS REPEAT

If I had thought Matt's revelation was a one-off, I was mistaken. By the time 2004 swung round, Matt and James weren't achieving the performances they expected and, in the final six months before the Olympics in Athens, there was a last-minute switch to convert to the coxless fours. Steve Williams and Alex Partridge were drafted in to complete the crew.

Sadly, Alex suffered a punctured lung a few weeks before the Olympics and Ed Coode (more on his redemption in a bit) replaced Alex at the eleventh hour. The Athens final was as excruciatingly close as the 2001 Worlds in Lucerne had been. This time, however, the British crew led all the way but were hounded and badgered by the World champion Canadian crew. The Brits triumphed, but only just. It was a monumental battle between these heavyweights (literally); one that would develop deep respect between the individual rowers and from journalists, commentators, athletes, coaches and sports fans alike.

In Athens, I was working at the GB Athlete Lodge in a dual role of running the centre as well as running the acclimatisation programme for the Games. It served as a central base from which I could distribute ice jackets and specially developed rehydration drinks, and provide advice to anybody struggling with pollution.

To my surprise, Matt shared his inner thinking with me once more when I saw him at the lodge the day after their bum-squeakingly tight race.

"I thought about you again in the final yesterday."

This time, exasperated, humbled and still incredibly surprised, all I could come up with was some self-deprecation and humour.

"Oh God, Matt. What are you doing? Will you please stop thinking about me during races? Will you please just take the lesson and unleash it on those oars, and take me out of the equation? You'll probably find a hundred more watts if you haven't got me in your mind!"

The lesson was clear to me. I needed to ensure that any interaction with an athlete, coach – or any other human, for that matter – that lacked anything other than delicate, responsive, wide-ranging and interconnected consideration would be substandard and could have a negative effect on someone else's efforts, motivation and inspiration. If I was capable of influencing others, I was certain to the nth degree that others could be much more influential but might not have realised it, so were probably not maximising their influence.

The lesson I learned in the summer of 2001 was that I needed to become better friends with tact, intrapersonal reflection, conversational oratory and timing. If I had an idea to share, I needed to be careful when, where and how I launched it. The lessons learned that summer propelled forward the importance of personal development. From that moment on, I committed to an even deeper level of personal development. I vowed to do more to support others who might find themselves in a similar situation.

I think the words of the Chinese proverb are most relevant to this realisation:

Be careful of your thoughts, for your thoughts become your words.

Be careful of your words, for your words become your actions.

Be careful of your actions, for your actions become your habits.

Be careful of your habits, for your habits become your character.

Be careful of your character, for your character becomes your destiny.

In Silvretta, I had indeed been a little bit saucy, edgy and perhaps even interrogative. I had an idea that I thought was relevant, and I had pushed that idea as far as it could go before I became insistent, but I had no sense of the potential influence that sharing my thoughts could have. At the time, I was reflecting on whether the timing was right, whether it was too aggressive, whether it was as tactfully presented as it possibly could have been, and whether our relationship was dented because of my proposition.

This case example gave me clear lessons in:

- Taking care in handling thoughts, words and actions as they can influence and nudge in ways I couldn't have imagined.

- The need to share my questions, ideas and thoughts with others before I launch them on coaches and athletes.

- The fact that progressing talent is a delicate balance between doing the basics well and pushing for innovation.

- Appreciating that high-performance sport is like the combination of a tightrope and a rollercoaster.

CHAPTER 7: TIGHTROPE ROLLERCOASTER

"Do nothing, say nothing, and be nothing, and you will never be criticised."

Elbert Hubbard, *in John North Willys*

THE INJURED KUDU

The kudu, a species of antelope[xxii], has evolved to possess long legs. Its limb levers needed to benefit its abilities and requirements to survive and breed as a species on the plains of Africa. Long lever length has greatly helped its ability to survive by ensuring the muscle force generated from the hips, shoulders, elbows, knees and hooves is transferred to a high-speed gallop. Over time, the longer-limbed kudu developed greater speed and avoided becoming prey. It passed on its genes for longer limb length, while its slower, stubby-legged siblings would

---

[xxii] *Yes, that's right folks, I've changed pace. This chapter's different.*

become food. Evolving longer limbs gave it greater speed and a better chance of survival.

However, longer limbs come at a cost as there is a greater propensity for injury. An injured kudu, or a kudu that becomes injured in a hunter pursuit, is unlikely to survive. The slow, short-limbed kudu might be vulnerable to being caught, but is less likely to get injured. If the stunted kudu lacks for speed, it needn't worry about being caught so long as there are a few, fast but delicate, kudu at the back limping around, because these will be the ones that are picked off. For a species to benefit from a characteristic, it must balance a happy medium of risk and reward.

Figure 28. The kudu antelope, with limbs evolved to balance speed and resistance to injury.

What is life like for the kudu on the planes? If you could hear some of the chatter in the herd, I would imagine it would go something like this:

"Cor blimey[xxiii], there are a lot of lions around today. That's three attacks in one day. The first one we all got away, the second one we lost a few, the third one we all got away again. It's a bit of rollercoaster, isn't it?"

"Bless Jim, he just isn't fast enough. I don't hold out much hope for him. I think he'll get nabbed at the back of the chase! It's a shame about Jane being injured. She's normally so fast with her long limbs; she's normally the first away and safe. Janice has it just about right. She's fast enough and robust. It's a bit of a tightrope, isn't it?"

I was once in a workshop about sprinting, and inevitably the topic swung round to the recurrence of hamstring injuries. A young, ballsy coach piped up, boasting, "My sprinters don't get hamstring injuries. My training is designed so carefully they don't ever get them."

For sprinters, this is a very bold claim. The young coach might well have constructed a brilliant programme to prevent hamstring problems. It is possible. However, a wiser coach delivered a killer putdown: "They don't get any hamstring injuries because they're not fast enough!"

In this respect, the young coach might have been prioritising hamstring injury prevention to the extent that other high-speed running characteristics were not being trained adequately. Alternatively, the hamstrings were not being put under the extreme strain experienced by top sprinters because the athletes did not have the natural ability to sprint at a world-class level. Either way, until the young coach had an international medallist on his hands, the validity of his claim would be hamstrung (see what I did there) by this limitation.

This example typifies a constant discussion about the training of an elite athlete relating to the number of days

---

[xxiii] Most experts agree the kudu would adopt a cockney accent if it achieved full speech abilities. Even if experts don't, I do.

lost due to injury and illness. For example, in the lead-up to the London 2012 Olympics, TeamGB lost a total of eight thousand and thirty days. That's a total of twenty-two years including injury (seventeen years) and illness (five years). This figure was rightly used to illustrate the potential of improving athlete availability to train, and for competition preparation.

There is one sure-fire way to ensure a poor medal count, and that is to leave your best athletes at home because they are unable to compete due to injuries or illnesses. Equally, training progression will not be as smooth or reach the same heights if an athlete is dogged by persistent injury or debilitating illness.

However, the ambition should not necessarily be to completely eradicate the number of days lost to injury or illness. It's easy to design a training programme for an athlete if the objective is to minimise injury or illness. The programme would be soft, manageable, enjoyable and safe – involving twenty-minute easy stationary cycling, for example – but it almost certainly wouldn't propel them onto the podium.

An athlete, and therefore the coach who designs the programme, must test the limits of what an athlete can do. If they don't, they can never be sure they have exacted all the potential from their athleticism. Quite simply, they might have gone faster if the training had been more rigorous. Thus, the impossible equation must be balanced.

If an athlete gets injured or ill, they have trained[xxiv] too much. If they never get injured or ill, they haven't trained hard enough. An athlete's career is certain to contain

---

xxiv *Or it is due to inadequate recovery as denoted by the shift from 'overtraining' (we think there is too much training) to 'under-recovery' (we don't think there is enough recovery) and then to 'unexplained underperformance syndrome'? (Oh, for goodness' sake, we don't know what it is any more.)*

injury and illness setbacks. This is the tightrope they continually tread: doing enough, but not too much.

In competition, probability is set even further against them. If any given athlete is selected to represent their country at the Olympics or Paralympics, odds-on they won't bring home a medal. There are currently three hundred and two events at the Olympics. Multiply this figure by three for each of the medals awarded, giving a total of nine hundred and six medals. Approximately ten thousand five hundred athletes attend the Olympics. Even when you allow for multiple people per medal, such as the hockey team, the odds are very long. The odds are better at the Paralympics, with five hundred and three events held and four thousand three hundred athletes competing, but most people will still lose.

Over the course of an athlete's career, even the talismanic greats such as Usain Bolt, Michael Phelps, Jesse Owens and Fanny Blankers-Koen suffered numerous losses. Granted, these particular champions will win more medals than most, but they will all vividly know the discomfort of defeat. All athletes do, and it is an inherent, predictable and clinical fact that if athletes want to succeed they have to experience defeat.

ED

Back to rowing. Sitting in the stands at 10.00am on September 23rd 2000, you would naturally assume our focus was on the men's coxless fours – the Redgrave, Pinsent, Foster and Cracknell boat – but the coxless pairs with Greg Searle and Ed Coode was about to take place. Greg had been Olympic champion eight years earlier and Ed had seamlessly substituted for Tim Foster into the coxless fours the previous year at the 1999 World

Championships, winning his first world title in the process. Ed's progression had been swift through the late 1990s and he was genuine competition for Tim's seat. In the background, throughout the 1999-00 season, Tim and Ed would vie for coxless fours selection. Tim finally got the nod following the spring trials.

Figure 29. Ed Coode and Greg Searle undertake final preparations for the Sydney Olympic Games at the Hinze Dam in the Gold Coast.

Greg and Ed formed a very strong pair. They had made an indifferent start to the season, but coming into the Games they were in excellent shape and were determined to make an impact. They won their heat convincingly, placed second in their semi-final and looked set for a medal.

As I sat in the stands surrounded by the Pinsent, Searle, Redgrave, Cracknell and Coode families, everybody in the British quarter was 100% behind Greg and Ed when the race began. They led for approximately 1,750m and went into the final eighth of the race heading for the gold medal. But, stroke by stroke, torturously, clinically and disastrously for the guys in the boat, they fell from first to fourth in a matter of seconds, losing a medal on the line.

The atmosphere in the stands was one of painful defeat and disappointment for the people we had worked with, and for their family members and friends. I later saw Greg and Ed undertake a painfully emotional interview with Steve Rider. They were subjected to inane and repetitive questions that raked over their defeat, such as, "Can you describe your feelings in the final stages of the race?"

Just fifteen minutes later, the coxless fours race was about to start. There was a great sense of anticipation in the air. This was the moment the crowd, the British team and the four million people staying up until midnight back in the UK had been waiting for. If a gold medal was going to come from anywhere in TeamGB, it was expected to come from the coxless fours.

I was primed and twitching. Our nerves had had about ten minutes to recover from the previous race, but as this next race approached, up went the palm sweating and heart rate. Two minutes before they were due to start, Ed Coode came up to the stands and was edging along the row to take the spare seat beside me.

I obviously felt sympathetic towards him, but I was also hugely aware of what was about to take place on the water. As we sat there, the coxless fours crew was about to win a gold medal. Ed had been an oar's blade from being in that boat, so we would have expected him to have headed for gold with them.

Before I could say anything, Ed shuffled past a couple of spectators and started rummaging around in his bag. He took out his heart rate monitor and watch, and handed them to me.

"Steve, here's your heart rate monitor back," he said.

I can only imagine he had experienced the same car-crash inducing radio-adjusting befuddlement as I had in Lucerne.

"Oh my God, Ed, you can have it! I'm so sorry, mate. Are you okay?" I was wobbling on the tightrope.

I wish now that I had had the mental agility, optimism and inspiration to have phrased it more tactfully: "I'm so sorry for you, Ed. You deserved a medal today. But you keep that heart rate monitor, because when you get over today, you're going to start training again and in four years' time you'll be in Athens to collect your Olympic gold medal."

I would have needed a touch of crystal ball mysticism to know that Ed would indeed collect a gold medal at the 2004 Athens final with Matthew Pinsent, James Cracknell and Steve Williams.

The coxless fours set off. At 500m they were leading and controlling the race, but I could feel the heat of discomfort in my mind. Ed was experiencing one of the deepest nadirs in sport. Plucky, happy 'also-rans' feel pain on losing, but it is not as deep as having victory plucked away at the last minute. Favourites who don't win, especially those who haven't won a major, feel the deepest pain. Ed should have won something, but sadly they went for it and returned home empty-handed.

Somewhere between the 500m and 1,000m mark of the coxless fours race, Ed realised he'd had enough. I presume he was overwhelmed by the emotion of his loss and from the impending win he was nearly part of, which made the race too unbearable for him to watch. He got up with the heart rate monitor still in his hand and took himself off into the crowd.

Peter Keen, who is a legend of British sport, architect of the 'no compromise' approach that escalated British sport up the medal tables, performance director of British cycling, and coach of cycling great, Chris Boardman, tells a tale that typifies the sense of doom-riddled anticipation in competition. Chris Boardman had been lining up for a World Championship final in the individual pursuit. It was Pete's job to hold the bike steady before the gun fires and the cyclists blast away from the start.

In the midst of the competition, the nerves, the self-doubt, the competitors and the inevitability of results, Chris turned to Pete, his long-time trusted coach and questioned: "Why does it always have to feel so shit?"

This sums up the emotion, self-doubt and misery many feel immediately before a competition. There are often attempts to buoy confidence in the pit of apprehension as the competition looms.

I teetered along the tightrope with Ed. Perhaps I should have followed him to check he was all right. Perhaps I should have shown greater care and consideration for him. However, the crowd was erupting around me by this point and I was about to see the first people I had supported win an Olympic gold medal. Selfishly, I stayed and cheered. I let the full emotions of the surrounding family and supporters take hold and I celebrated their win. However, my elation was muted by the thought of Ed and the realisation that everybody who has ever tried and failed would be feeling the same way.

The following day, I willed the men's eight to their Olympic gold medal. In my memory, I cheered them all the way down the course, although the video recording I made of the final shows my shaky hand accompanied by silence. However, with about 200m to go, overwhelmed with the hope that there would be no repeat of an Ed and Greg style swallowing up from first down to fourth, I simply shouted one word: "EVERYTHING!"

The final few seconds of the race unravelled with victory. One day earlier at the same stage of the race, victory had unravelled for another crew and defeat had closed in on the Brits.

A Bolt from the Blues

As I got on the bus at Lake Penrith in the early afternoon of September 24th 2000, I reflected deeply on the intensity of my first full Olympic experience. With a torrent of images, emotions and sounds running through my head – the four, the eight, Ed, Greg, the women's first medal (quadruple sculls clinched a silver), the friends, the families, the thank yous and the tiredness – questions started to bubble in my brain, 'How did that happen? Did that really happen? Did I contribute? Was I just fortunate? What is this all for? Have my efforts and the efforts of the coaches and athletes achieved anything of meaning? How is sport relevant to a world full of poverty, inequality and atrocities?'

I was acutely aware that I should have been overwhelmed with joy, which I was. However, I was surprised to find myself asking deeper questions. My head felt like a washing machine with the good, the bad and the ugly all churning around at the same time. The rollercoaster was up to full speed. That night, I attended the after-competition party with the athletes. It's amazing how many free drinks get handed out when you're with people who have Olympic gold medals around their necks!

When I returned home to the UK, I played the hours of videotape my wife Rachel had recorded for me. When you're at an Olympic Games, you don't see an awful lot of coverage because you're too busy. This is especially true when you're overseas and the host nation is blinkered to the exploits of their own national team[xxv].

Having spent upwards of twenty hours combing through the Olympic coverage, playing the races over and over, watching the interviews and reading through the papers, I found myself in a phase that I have since termed 'jubilant mourning'. I felt hugely privileged to have been part of this team and fortunate to be doing this type of work. For the first time since 1968, Britain had finished in the top ten of a non-boycotted Olympic medal table and I had been part of the mission. Yet all the questions I had asked myself on the bus back from Penrith and many more were still being unpicked, one by one, over and over, as I ruminated and questioned and reached for meaning.

As I began to resurface and talk to other people, I began to realise this was a normal phase, known as the post-Olympic blues. It's almost unavoidable. You know it's coming, and you can plan for it, take counsel from other people, book a nice holiday and plan to do something exciting, but sure enough it will hit you. It just depends how hard it hits.

The rollercoaster will lurch up and down, left and right when you're working with talent. My experiences in Sydney would be replicated over and over again throughout my career. Having worked with Hayley Tullett from 2002 up to the World Championship medal in Paris in 2003, she approached the Athens Games as a real contender. The struggle through her heat and failure to qualify beyond the semi-final was painful to watch. We watched the 1,500m final with a clear tinge of what might have been for Hayley. Her best performances would have brought her close to a medal.

[xxv] *The good old BBC mostly shows British athletes, but does an excellent job of showing good sporting coverage, irrespective of who is competing.*

The 1,500m final in Athens was laced with glorious atonement for Kelly Holmes, who had spent nearly her whole career trying to capture and achieve her true potential. Her career reads as a continual struggle with injury and occasional windows of health where her performance ignited and reflected her capability. I knew Kelly Holmes, but not nearly as well as I knew Hayley. I was, along with the whole nation, ardently pleased that Kelly's persistence was rewarded with 800m and 1,500m gold medals, but understandably, as part of Hayley's team, I wanted success for her, not least because I stood (well and truly on the tightrope) right next to her as we watched the 1,500m final. The rollercoaster continued to rise and fall.

As the final heptathlon javelins were thrown in the Bird's Nest Stadium during the Beijing Games in 2008, I knew Kelly Sotherton's quest for a medal (on the day, that is, see chapter five for more on the retrospective award nine years later) was over. She would have needed to throw an incredible personal best in the javelin or to beat her nearest rivals by five seconds in the 800m event, and that just wasn't going to happen.

Once the javelin event was over, I met up with my great friend, Andy Allford, who had managed to get hold of a premium ticket for one of the best seats in the stadium, about 50m into the home straight. He was positioned right next to the VIP area[xxvi]. With some assertive waving of my GB accreditation, anxiously stating, "I'm a doctor, I'm a doctor" at the disconcerted guard (he didn't need to know my doctorate was in philosophy) at the stand entrance, I joined him in his luxury seat.

[xxvi] *It's amazing which tickets are available if you're in the right place at the right time. Always make friends with the corporate hospitality team. Always.*

Figure 30. Kelly Sotherton under the intense spotlight of the Olympic Games in 2008.

From this privileged perch, I watched Kelly Sotherton's 800m. She valiantly strained for every second and heptathlon point she could, setting a new personal best for the event. The rollercoaster would undulate again. I was proud of her efforts. As her running coach, I was pleased with her new personal best. However, I was naturally hugely disappointed for her, for the coaching team and for TeamGB that she didn't medal.

Twenty-eight minutes later, after the 800m was over and having applauded the traditional convivial procession of heptathletes, the starting gun fired for the 100m. All evening, while I was preoccupied with Kelly's exploits, huge anticipation was building in the stadium around a certain Usain Bolt, who had rocked sprinting, athletics and the world of sport that year. Bolt had taken the 100m world record in only his fifth senior 100m race.

His casual, nonchalant, laidback and genial style made him an instant crowd favourite and incredibly exciting to watch. The sense of anticipation sizzling around the stadium built inexorably toward the final. The crowd was taking it in turns to point him out. It appeared as though all the photographers had their lenses trained on Bolt, taking picture after picture as flashbulb after flashbulb sparked. Every one of Bolt's gestures, which so famously broke the ice for all athletes to do something silly and just chill out pre-race – his salutes, smiles and winks – was met with amazed laughter. It was as though there was no else in the race (pop quiz: who came second and third?[xxvii]).

Bolt's overt displays, waves, dance moves and of course the lightning bolt pose had the same effect. The 100m final is the blue riband of blue ribands, and he had somehow turned a prowling, simmering pre-event ritual (Linford Christie would often growl at fellow athletes) into a carnival atmosphere. It was infectious.

As the gun went off, Andy I watched Bolt rip the Beijing track to shreds in front of us with his chest thumping, lip biting and showboating, high-kneed stride. It would be no exaggeration to say that he left the most electric aftershock reaction. There were cheers, but these were dampened by the need of the crowd to collectively shout "WOAH!" Everybody was turning to each other to check what they had seen. "What on earth was that?" they cried. Destruction of a world record in the premier Olympic event was one thing, but to do it so casually, so incidentally, running half of it like a footballer's goal celebration, shook the crowd.

The electricity continued to buzz long after the event. His lap of honour was in danger of becoming a decent 400m, such was his enormous speed. But he managed to milk his performance to high heaven. Everyone wanted a piece of Bolt. We wanted to see him again to check that he

---

[xxvii] *Richard Thompson, Trinidad and Tobago took silver. Walter Dix, USA took bronze.*

wasn't an alien. We all knew we had witnessed perhaps the most incredible moment in Olympic history.

Figure 31. Usain Bolt celebrates winning the 2008 Olympic title with a new World Record in Beijing.

The prominence of the 100m is unquestioned, which makes any win significant. A world record is rare, especially in a final, as everyone is normally too tense to run as efficiently as possible. But not Bolt. The margin of his win was freakishly apparent by the pictures showing the normal race going on behind him. He was of a different ilk; no one had been seen to run like this before.

On top of all that, he did it in style, showing the world he could run like no other. All this added up to it becoming recognised as one of the foremost moments in Olympic and sporting history. The emphatic nature of Bolt's performance was such that the immediate aftershock was met with a reality-infused wish: "I hope he's clean." It's a disappointing state of affairs that, when a performance is so spectacular, audiences have learned to admire and doubt at the same time.

However, I recognised the significance of that moment back then. I knew that, as a sports fan and slightly nerdy follower of the 100m, I was truly fortunate to have watched that moment. As a former (albeit rather rubbish) sprinter, it was truly humbling to see such a performance. I truly savoured the moment. The rollercoaster was at one of its highest heights, having only previously been so high watching Redgrave et al, the men's 8+ and Cathy Freeman beforehand and subsequently watching Jess Ennis-Hill. But at the back of my mind was Kelly Sotherton, her feelings and how her event had previously unfolded. This was a completely different dilemma from the Kelly versus Jess loyalty pull in 2007. It was a juxtaposition of confusion between feeling elated and disappointed just minutes apart.

Kelly is a tough cookie. If she had been there to hear my inner voice[xxviii], she no doubt would have said, "Don't be so soppy! Forget me and enjoy watching Bolt!" While Kelly's on-the-day disappointment didn't diminish the 'flashbulb' experience of the Bolt sensation, it perfectly illustrates the gravitational pressures as the ride dips to a nadir followed by a rapid ascent into the lofty weightlessness at the zenith of the ride.

These highs and lows are common, and they are felt by many. However, a new rhythm has arisen in the modern-day rollercoaster; that of the retrospective doping ban/medal upgrade. This is laced with a melee of emotions ranging from joy, justice and compensation through to betrayal, bitterness and deprivation. If you are an athlete or work to support an athlete, the ride might careen in this way at any given point, either during the

---

xxviii *Not good for anyone's well-being!*

athlete's career or even after it is over. Practitioners should expect many of these twists and prepare for them.

REALISING POTENTIAL

Not all highs and lows are so severe and sharp. The undulations can come in different disguises on different occasions and under different circumstances. I can still recall that, from the moment I first started to work with Jessica Ennis-Hill, there was a real sense of anticipation about what might lie ahead. Jess had such overt potential, coupled with a great sense of awareness, critical thought and a no-nonsense approach. I knew she would be able to handle her own talent. Her very potential was enormously exciting. And as her career unfolded, the excitement built as this potential was realised.

Throughout my career, I have been equally convinced by the potential of other athletes who didn't realise the heights they perhaps could have done. One athlete I worked with for five years or so, who I was convinced was 'the chosen one', was Laura Finucane.

I remain convinced that Laura had all the necessary hardware to win at the highest level. She had a tremendous work ethic and a huge appetite for hard training. I have rarely seen someone so capable of digging themselves into such a deep and dark hole of pain in training. Steve Williams and James Cracknell were similar. The rarity Laura possessed was the ability to flog herself in repetition sessions. She would be at the very ragged edge of fatigue, but then somehow she would always be able to drive herself to deliver a further gigantic effort and, in the process, she would completely bury herself in a whole world of metabolic distress.

Figure 32. Laura Finucane, one of the most talented athletes I have ever worked with. Her mechanical structure was one of her greatest strengths, but also proved to be her weakness. (Reproduced with permission from Job King, copyrighted.)

She had a great engine, but her other distinguishing strength was probably her weakness. Laura had long levers, which made for elegant running, giving her greater speed for the same energetic effort. Laura's undoing was akin to the kudu speed balancing act. Laura's levers would gift her a great stride length and, with it, speed. However, they were especially susceptible to injury. Laura had to adopt a specially adapted training programme to stay injury-free. It was a constant iterative challenge we had to work on.

Even then, it wasn't enough. Laura eventually had to retire early because of persistent and recurrent injury. It was hugely disappointing that she didn't make it. It was sad for Laura that she was hampered by her own mechanical structure, which was built so well for speed but proved too frail for the continued, progressive training required to hone her abilities and realise her potential. This is the cold natural selection process of sport. This is a different type of low; a sidewinding twist in the rollercoaster.

THE EFFECTS OF ALTITUDE

In Jon Krakauer's epic account of the disaster on Mount Everest in 1996 *Into Thin Air*[7], which was dramatised in the film *Everest*, there is a mountaineer by the name of Doug Hansen who was scaling Everest for the second time. He had climbed it a few years earlier with the same guide, Rob Hall, when the pair had been forced to turn back less than 100m from the summit owing to bad weather.

Hansen was highly motivated to reach the summit in May 1996. He had effectively divorced from his wife arising from his affair with mountaineering. Equally, Hall felt the pressure of not summiting Hansen on the previous occasion, as described by Krakauer: "It would have been especially hard for him to deny Hansen the summit the second time." Krakauer's tale tells of the desperate ambition Hansen displayed.

When the storm struck, this confluence of factors led Hansen and Hall to make a series of mistakes as they pushed to the summit, despite it being much later than the agreed turnaround time. When he realised the summit might not be possible, Hansen had a choice to either turn back once again or risk his life in continuing. Hansen was resolutely determined to reach the summit this time and paid the ultimate cost, as did his trusted guide, Hall.

When expeditionary mountaineers pursue such summits, they are extraordinarily exposed on the tightrope they tread. Hansen's case shows the fragility with which risk and reward are weighed. It shows us how, in a parallel world, the balance between life and death is the difference between making a simple decision; a decision that on one day could work just fine and on another could lead to loss of life. A hair's breadth separates the mountaineer from success or disaster.

Mountaineers won't jump into ascending one of the big 'eight-thousander' (greater than 8,000m) mountains straight off. They will learn their technical craft and build up their progress steadily to ensure that their skills match the difficulty of the challenge. They will plan their climbs meticulously to make sure they have the necessary equipment and understanding of their routes, and that they are physically and mentally prepared. Solo adventures are the domain of the extreme experts. The clear majority of expeditionary efforts are undertaken in teams, pairs or guided by an experienced mountaineer. In this sense, the athletes are the mountaineers, the coaches are the guides and the support scientists are probably the Sherpas.

As the mountaineer climbs further and further upwards, the air thins more and more. Physically, the tissues scream out afferent feedback, signalling a requirement for adequate oxygen. The body inflames with oedema as the partial pressure of the gases send the body into a distressed state of shock. The lack of oxygen in the brain causes confusion, delirium and hallucinations.

The mountaineer is in a rarefied position in which judgement can be impaired. The primary responsibility of the guide and the Sherpa is the safety of the person in their charge. They must give the best advice available while continually keeping the end-goal achievement of summiting in mind. They tread a difficult tightrope in developing their reputation as a guide or Sherpa who can get people to the top, but who can act decisively and with authority to withdraw if there is too much risk.

As athletes get closer to a big competition, they crave confidence as a mountaineer craves oxygen. They are constantly looking for certainty about their physical abilities and the probabilities of achieving their goals. Any decisions made at the last minute when the athlete enters 'thin air' should be tried and trusted. Like a mountaineer, they must look into the eyes of the guide or Sherpa and be able to trust their judgements implicitly. They will constantly evaluate what they are told and crosscheck against their own perceptions and judgements. Critically, it is the behaviours of the support staff that will speak loudest at such high altitudes.

The unnerved coach or support scientist, flapping about, exhibiting stress and showing a lack of control, will undermine confidence, drawing oxygen away from the athlete at a critical moment in the climb. This is the tightrope support staff and coaches walk. They must give the athlete just about enough, or ever so slightly more confidence than their abilities are due.

Any more than this and the athlete will lose confidence in the coach and support team because there is a whiff that they have drifted out of touch with a valid reality. Any less and the athlete will overly question whether they can do it. With a touch more confidence than their abilities match up to, their confidence is more likely to be calibrated at just the right level when they are on the line and the questions come.

As I undertook my undergraduate studies studying the basics of sports performance, especially the psychology part, we had it drummed into us that athletes must have a huge amount of self-belief to train and perform at their optimum. As I got to know some Olympic-level athletes, I was surprised to hear their constant questioning, obsessive perfectionism and need of assurance. Where I was expecting Teflon-coated, self-confident athletes, the extreme conscientiousness and focus on mastery propels athletes to want more[8]. I have only met a few athletes whose self-confidence far outstrips their abilities.

If athletes are overconfident, they are less likely to question their status, and if they are not questioning their status they are probably not yearning enough for development. When athletes have this obsessive perfectionism coupled with the selfishness to make critical decisions and the wherewithal to question wisely, they appear to have the best chance of maximising their physical capabilities.

However, delivering on the day requires a constitution that is able to disconnect from the emotions and that can send the mind and body into autopilot under the stress and pressure. Coaches and support staff must do everything in their power, through their words, expressions and body language, to ensure that nothing destabilises an athlete's ability to perform. Doing so is tantamount to playing practical jokes at 8,700m.

ATTRIBUTES

When NASA advertised for potential aerospace pilots in 1958, the job description listed the characteristics required. They asked for men (how sexist) of about 5'11" (180cm) in stature, aged between twenty-five and forty, who possessed a science degree, high intelligence, high mathematical ability and experience as a jet test pilot. There were just six technical requirements.

Then it listed a whole host of personal requirements. The applicant was to be healthy, to have the ability to command, to have the ability to follow orders, to be motivated, creative, adaptable, sociable, mature, decent, and psychologically stable, to have experience in dangerous and stressful situations, and my absolute favourite, to have the ability to sit still in a dangerous situation.

All these attributes were expected to be tested in the unknown ventures of space exploration and the potential lunar landing. The candidate would need to use his own personal disposition to ameliorate the stresses and to keep things normal. These men were expected to perform under pressure and to dampen the rollercoaster's highs and lows.

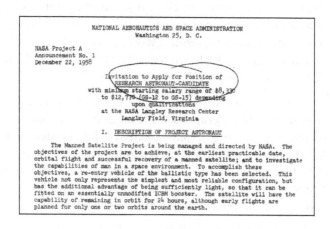

Figure 33. Copy of the original NASA job description for astronauts to undertake manned space flight.

During the lunar landing operation – amid the error codes, fuel warnings, visual information from the lunar module's window and judgements about the unsuitability of the lunar surface – Armstrong switched to manual control and decided to apply the engines to find a better landing area, not knowing whether they would find one. During this time, the onboard monitoring system recorded the jump of his heart rate from seventy-seven to a hundred and fifty-six beats per minute when they finally touched down in the Sea of Tranquillity with eighteen seconds of fuel left in the tanks.

This is an epic example of performing under pressure. While Armstrong and Aldrin undoubtedly felt the anxiety of the situation, the lesson we can learn is that, impressively, they had drilled the landing scenario over and over, up to a position of having the absolute minimum ten seconds of fuel remaining at the moment of touchdown. They had rehearsed the potential eventualities, meaning they were part-acclimatised to the situation that arose. As the lunar landing unfolded, the rehearsed plan placed considerable psychophysiological stress on Armstrong (well it would, wouldn't it?), but just imagine how much more stress he would have experienced if he hadn't rehearsed.

A sports scientist or coach is not expected to perform anything near as grand as landing a lunar probe on the moon, nor is an applied scientist's job lauded with a tickertape homecoming. However, he or she must tread the tightrope of providing service in pressurised situations, and to deliver at the right moment in the right way, while the rollercoaster dips high and low.

The environment in which applied scientists or coach must work can vary hugely, and ultimately the pressure they are under is a manifestation of their own perception. However, the escalation of pressure will typically mount from: junior club to senior club; local club to regional team; national team to international team; domestic training to training camp; day-to-day training to pre-competition; holding camp to competition; and World Championships to Olympic Games. Practitioners are best served if they spend time acclimatising to each level, developing their skills and preparing for the next level of intensity, knowing that their skills will be tested and their behaviours will be exposed at a new level when the pressure ratchets up as the competition importance increases.

As a support practitioner, the key relationships are with a coach, an athlete and other practitioners in a multidisciplinary team. Therefore, you must be able to project yourself and dance along various tightropes:

- Team worker but independent
- Interpersonal but intrapersonal
- Tolerant but critical
- Deferential but assured
- Openminded but discerning
- Patient but persistent
- Humble but assertive

If practitioners can balance these factors and more, they will be tightrope walking to success. The reality is, practitioners need to develop this versatility through seeking, receiving and digesting feedback, through adaptability, through trial and error, effort and reflection, through loops of matured but fast-tracked learning.

TRIBAL BASTIONS

Athletes are expeditionary mountaineers. They are prepared to put their bodies through extreme conditions. They are prepared to test their minds to boiling point. They are the representatives of our tribes.

In just a blip of human history ago, they would have been the tribal representatives who took up the lonely mantle of being prepared to defend their territory. They would have been prepared to put their lives in danger to hunt prey. They would have been the brave people who set out to conquer new lands. That is why we revere them when they return with the medal of success; equivalent to the prize of food or the news of fertile pastures. That is why we uphold these bastions, and it is why we treat them with care, because when they are exposed beyond the tribal boundary they too will be questioning themselves, their ability and their wherewithal to deliver success or handle failure.

The day-to-day travails the kudu face on the planes are similar to those our tribes faced not too long ago. It is hard-written into our behavioural and social needs for us to test our individual and collective capabilities. Do something too risky and you might fail, but if you don't do enough, the risk is tribal suffering. If I look back over my experiences of working with the elite, I have seen and felt that:

- There is a delicate balancing act of how much athletes push themselves and how much coaches and practitioners stick with the basics or twist with experimentation.

- There will be ups and downs on a day-to-day basis, and they need to be planned for and coped with in order for the athlete to thrive in the sphere of high performance.

- Sport would be an empty place if it wasn't enriched with the challenge of treading the tightrope and riding the rollercoaster.

Chapter 8: All True to Altruism?

"Commerce, trade and exchange make other people more valuable alive than dead, and mean that people try to anticipate what the other guy needs and wants. It engages the mechanisms of reciprocal altruism, as the evolutionary biologists call it, as opposed to raw dominance."

Steven Pinker

Supporting a champion

When I finished my undergraduate studies, I was at the height of my Dunning-Kruger spike of narcissism. I had my first-class honours degree (only just), had finished my studies with a flurry of good marks and thought I knew everything. Looking back, boy was I confident then; cocksure even. As I have previously described, I had accumulated a good deal of experience of working with people before I had reached the point of hearing Redgrave's Exocet.

When I undertook my A-level studies, I would get together with a group of friends and work out ways in which we could test our new knowledge with each other. This involved shenanigans such as running up and down the zigzag cliff paths on the front at Bournemouth beach to experience maximum heart rate, ventilatory threshold, dyspnoea and redistribution of blood flow to the muscles.[xxix]

Following anatomy classes, we would focus on a set of muscles each week and flog them in the gym, knowing the muscle soreness that followed would leave an indelible memory of the muscle group and what it does. My friends, Adam Riccio and Mark Robinson, and I set up the sixth-form college gym when we were seventeen, and would put anyone willing to walk in through the door through their paces so we could see how well the theory stacked up in real life. During my undergraduate studies, I worked in various gyms and actively sought out athletes who were openminded enough to consider receiving some advice. By the time I had finished my undergraduate degree, I had a bank of experiences from five years' worth of work that I'm sure stood me in good stead to get my first job in performance sport.

That first job was with the West Midlands Sports Council, based in Worcester. At the time, I was one of only ten full-time sports scientists in Great Britain. As of today, that number exceeds five hundred in the UK institutes of sport alone. Taking into consideration all the professional clubs, university providers and independent professionals, I have no doubt the actual figure is greater than three thousand in the UK alone.

As I took up my post as a sports performance officer, I could feel the responsibility of the position shaking off my

---

[xxix] *Including away from the gut, which often led to exercise-induced vomiting, which wasn't always pleasant for the tourists.*

egotistical narcissism by the second. After a day and a half of inductions from my boss at the time, Dr Andy Miles, he had perhaps had enough of talking and sat me down for some work. He presented me with the dataset from their recent physiological testing of the West Midlands rowing team. It was a box of dot-matrix printouts of expired air data readouts for twenty-five athletes on their programme. At the time, this was the height of new technology in physiology and it was my job to tabulate, organise and generally trudge through it to make sense of the information. For the applied scientist of today, this would take just a few fully automated minutes, ready for immediate feedback. Back in 1997, this was about five days' worth of work.

Matthew Beechey was the best athlete on the West Midlands programme, and I would later go on to work with him in the senior international team. As Matthew was studying at the University of Worcester, we ran his physiological testing. I vividly remember looking over at the box of data and, despite the prospect of crunching the information until my eyes turned square, I was full of excitement about getting to grips with measurements on funded athletes for the first time. I flicked through the concertina printouts to find Matthew's data and coo at some of his capacities. They were impressive; far higher than I had encountered in my voluntary support experience.

As I got to grips with Matthew's data, I began to sense the responsibility I had to him and to the work I would be doing to help him. While I had previously been highly considerate of the application of my knowledge with previous athletes, looking back I could see that it had been an environment in which I had been free to make mistakes and use the data as a test bed for my own performance. Now I was dealing with an international athlete and I had to get this right. This wasn't the place for me to be focusing on my own development and making mistakes willy-nilly, as that would prioritise my learning above his performance.

The ego I had felt most keenly at graduation was beginning to fall away. My interest and motivations were shifting towards helping others. I was beginning to realise that my job was about supporting other people. I understood that my ego could get in the way and that, if I displayed selfishness or acts of self-promotion, I would not be able to do the best job of developing the athlete. In turn, I would not be as successful in the role.

AN ALIEN'S VIEW OF SUPPORTING A CHAMPION

If an alien were to hover above any of the Olympic cities at the time of the Games and observe the sports teams in action, I wonder what its conclusion would be. It might be interested to see an advanced 'ape gathering' in purpose-built arenas. There would be noise erupting and bright lights associated with the occasional activities of a select few of the species, which might perk their interest. If the aliens were being objective about the amount of activity our species carried out at these gatherings, would they conclude that the main event was, in fact, the supporting cast?

If we were to peer down at the furious work of a colony of ants ferrying hardware, and shifting sand and soil, we might conclude that the main event was building a formicary nest. But the end goal is to protect and house the queen ant; to allow her to reproduce. The colony is a means to an end. The Games is the formicary nest and the exhaustive work involved in hosting, producing, planning and delivering one is for the cherished incandesce of competition.

Figure 34. The Olympic Games brings together the nations of the world for a massive sports day!

If you consider the amount of work the coaches, performance directors, support scientists, support medics, team managers, operations managers, transport staff, catering staff, security staff, hospitality staff, cleaning staff, planning staff and media staff (including journalists, commentators, camera operators and so on) do, athletes are vastly outnumbered by the supporting cast. However, these support staff are all there working to deliver the event; to deliver their team or their athletes.

Even then, the aliens wouldn't see the sports scientist or sports medic solely focusing on science and medicine. It would see them driving the bus, carrying bags, washing towels in the clinic, fetching ice, assembling gazebos and generally mucking in to ensure the athletes have everything they need.

It is completely paradoxical that we put so much effort into an event for which the outcomes and rewards are wholly arbitrary. The job as an applied scientist is to provide the best support to representatives of the tribe. My job is not to share the podium with the athletes or to grab the headlines and the glory. Nevertheless, the supporting teams are integrally involved in the athlete's performance.

Lord Paul Deighton, chief executive of the London Organising Committee of the Olympic and Paralympic Games (LOCOG), introduced a wonderful phrase to recognise the efforts of staff around the time of the Olympics and Paralympics. He called it the "glorification of the uninvolved". He was referring to the people who really had no input into the making of the Games or athlete performances, but were centre stage and ready to take the adulation associated with its successes. This always happens and continually irks, but does it really matter when you *know* in yourself that you have been involved, even if you get none of the acclaim?

The primary role of applied scientists is to support coaches in their pursuit of developing their athletes. Their clients are the coaches and athletes, and their place in the team is as background staff. They are the vocal coaches, the dance teachers, the choreographers, the musicians, the stage hands, the lighting technicians. But they are not part of the ensemble and are definitely not the main stars of the show.

However, an applied scientist can add percentage improvements to an athlete's performance. Perhaps, once upon a time, support scientists and medics were shy about claiming that they had improved performance, but this needn't be the case now. Contemporary applied sports science knows with confidence that it can improve performance within the rules, which is the most uplifting thing. That is about as bolshie as applied scientists will ever need to be, because our place is in the background.

WANT TO SUPPORT A CHAMPION?

Applied science and medicine is altruistic in character. In the classic Darwinian sense of the word, organisms function to maximise their 'inclusive fitness', which is measured by the survival and reproduction of the individual and the relatives who share their genes[9]. Sport can't claim to be a vehicle that promotes the survival of the species, although at a national level it is capable of spiking the birth rate[10]!

Indirectly, then, I suppose applied science and medicine is part of a movement that prompts the reproduction of Homo sapiens. This must be an evolutionary-level trigger that our tribal bastions are performing well, akin to a successful hunt or the seizure of new fertile territory, which activates the green light to perpetuate one's own genes.

As Boris Johnson, the Mayor of London in 2012, eulogised at the GB athletes' parade:

"And speaking as a spectator, you produced such paroxysms of tears and joy on the sofas of Britain that you probably not only inspired a generation but helped to create one as well."

Strictly speaking, altruistic behaviour is understood to be that which benefits others at a personal cost to the behaving individual. I can think of numerous coaches and support staff whose private relationships have paid a terminal price for their passionate pursuit of supporting athletes.

I recall one mentor of mine saying: "Steve, you can't have a successful personal relationship and a successful career in high-performance sport."

I fundamentally disagree with this statement. However, if you are signed up to a career that requires you to be away for a third or more of your year, travelling to warm weather training camps, altitude training camps, competitions, domestic work and overseas work, you will ultimately incur some personal cost to your home and personal life.

In 1998, I was so focused on ensuring that I didn't make mistakes while working with the rowing team that I admittedly put my personal relationships on hold. In the 1999-2000 season, I was away from home for more than a hundred and fifty days, and this took its toll. To weather this work demand, I had to have several, 'sit-down discussions' with my partner to discuss, negotiate and agree on what was acceptable and what wasn't.

I'm not sure I had a good night's sleep at all in 2008, so focused was I on getting it right for Kelly Sotherton and managing a complicated practitioner case. Many people probably wouldn't have accepted what I agreed to, while others would have given up even more in the name of sport. I truly believe you can have a successful personal relationship as well as a successful career in high-performance sport, but it is a fine line; another tightrope to be trodden.

I remember meeting up with some friends, a couple we had known for many years, on one occasion, and the chap asked me how things were going with work. I had just returned from two weeks away at a training camp. Having left at 4.00am to catch a taxi, a train, a flight, the London underground, a further train and then a final taxi, I was quite tired.

As I started moaning about the situation, I had to check myself because I had just spent two weeks in Formia, Italy. I had worked all of five hours per day, run on the beach every morning, watched Yelena Isinbayeva, the world record holder in the pole vault training every day and played basketball with World champion triple jumper Phillips Idowu. The accommodation and food had all been paid for and the temperatures had been at a cheeky 28°C every day, yet here I was moaning.

To make things even worse, the husband had just returned from a three-month military tour in Oman, undertaking night-time scouting missions over hostile territories in the Middle East. Throughout the trip his location had been confidential, even for his wife, and they only had contact if he called her out of the blue. You can guess how I felt! I was suitably embarrassed when he followed up with these details. In comparison, my complaints sounded as though I wanted utopia to be more utopic.

Working as an applied scientist in high-performance sport is a truly privileged position. Sports science is one of the most popular courses in Western societies, and working with elite athletes is one of the top-rated career destinations for such graduates. Every day, applied sports scientists get to work with the best athletes in the world in some of the nicest sporting venues in the world. What's not to like about that?

Indeed, there are personal costs, but it is a purposeful world in which to live. When I first joined a full-blown international team, the GB rowing team, it was clear to me that this wasn't going to be a nine-to-five job. Most days I was up at 5:30am and I wouldn't finish work until 11:00pm. I only had a couple of hours to myself during the day, and this time was normally spent processing data, calibrating machines or disposing of bodily fluids!

To operate in a world that accelerates talent, you must put the hard graft in, simply because you should be able to go toe-to-toe and work hand-in-glove with the coaches and athletes. Anything less and, trust me, the assurance you deliver will be shaken. The high level of expectation is unspoken, it is implicit. You will need to perform well, consistently and unflinchingly. So there is a personal cost, but it is one you must be motivated to take. Your motivation must be to help others.

However, applied science support is also a job. It is a long way from being a spontaneous, 'good Samaritan' act of kindness, care or giving. For many, this is a paid profession. So there is a personal gain for the person who supports the champion. This might appear to diminish the claim that applied scientists are involved in altruistic behaviour. I would imagine that purer altruists are more commonly drawn to professions such as nursing, counselling and clinical psychology; professions that require deep empathy, and derive much more purpose, happiness and contentment from serving others.

NEED TO SUPPORT A CHAMPION?

I am sure there is an egotistical aspect in all applied scientists. Many people who work as applied scientists in sport, including me, are failed athletes. Whether because of a lack of physical talent or mental strength, many haven't quite reached the highest heights. Most are highly motivated to succeed and are drawn to operate in high-performance sport. Perhaps in a 'pushy parent' type of way, we are living our dreams through the successes of the people we support.

Applied scientists are often quite open in admitting that they want to be the best at what they do. Many people in the profession want to be involved with the top athletes. Working with the best or being the best is admittedly highly alluring, intoxicating even, and what we do makes for great conversation. There are very few people who work in this world who could genuinely admit that they don't like the 'sexiness' of their work. Very few applied scientists could genuinely admit to not enjoying the status and standing of their roles.

Like others, I would have to admit that there is considerable personal gain to be had from working with champions. With a dash of irony, I admit that it's probably my willingness to support champions that has allowed me to progress through my career from practitioner to manager to leader. Where would I be without having worked with 'big names' such as Redgrave, Pinsent or Ennis-Hill? Even having the chutzpah to write this book arguably clashes with the humility required to be content in simply providing for others. I'm conflicted, confused and feeling thoroughly dichotomous just writing this! I suppose if I was truly humble and altruistic, I would be more like Gregor Mendel, founder of modern genetics, who went about his work monastically, carefully documenting his findings and filing his work away for somebody to find it after he passed away.

There is certainly an immediacy of today, with social media dominating the airwaves and brainwaves, which has created a generation that demands, "Look at me!

Recognise me!" This will only ever flare egotism and the need to live on your own terms in the future. My expectation is that if such ego rages in the applied scientists of the future, it will cause them to fail. Self-centred immediacy is a path toward the dark side. To give, to empathise, to be self-effacing and to be generous to others, is a path full of light.

A further definition of altruism is that you're giving to those in need, in danger or in distress. A president needs a bodyguard to take the ultimate risk so that he or she can to continue to govern as the leader of a country. A child rejected by its natural parents needs adoptive parents to give it food, love and care. The pursuit of becoming an elite athlete does not confer the same magnitude of basic human needs. The needs of an athlete are superficial, irrelevant and disproportionately smaller by comparison with the suffering that occurs in all quarters of our globe.

Elite athletes are, by definition, some of the most talented people on earth. Irrespective of their upbringing or the conditions of their childhood, their physical capabilities and mental aptitudes are privileged gifts that they must cherish, nurture and exploit if they are to be champions. They are highly privileged. These gifted, athletic, brilliant, often annoyingly good-looking freaks of the human species have such minor needs when compared with, the casualties of a hurricane disaster, migrants prepared to risk their lives to escape treachery in their homelands, the child who has to walk twenty-two miles a day to fetch water or the millions of victims of malaria, for example. So how do I expect you to accept my proposition that this is a group of people that is in need? And how do I expect you to accept my proposition that, to be effective as an applied scientist, you need high levels of altruism, empathy, care and generosity?

I don't expect you to. The world of high-performance sport is a glitzy, arbitrary, self-centred, glorified sports day. It is a demanding and intense world within which to operate. The application of science to the recipient – the athlete – does not fall into the category of a bodyguard taking a bullet for a president or adopting a child, where the cost might be your own life, or a great financial and pastoral cost.

Sport is a metaphor for human progress. It is symbolic of our struggle to continue to survive, exist and flourish. Altruism is often discussed in relation to acts that promote 'inclusive fitness' of the species. I have described our champions as our tribal bastions and representatives. If they can bring pride to a nation, they are still enacting the brave feats of tribal protection and propagation.

They are analogous to the hunters, bringing down prey so their offspring, cousins and elders can eat and continue to survive. They are analogous to the explorers who use their human endurance to survey new lands. They are the expeditionists who use their physical and mental fortitude to attempt to bridge the ravine or canoe across the estuary to unoccupied and fertile territories. Our tribe still celebrates when these champions return successfully from their exploits.

TOOLS TO SUPPORT A CHAMPION

Let's take the example of the first phase of stone tool development. The technology our ancestors used nearly three million years ago had probably been handed down to their forefathers, *Australopithecus afarensis*. I wager that there were three possible circumstantial routes to the origin of stone tool usage.

The first might be stumbling across a stone tool's functional benefit. Perhaps *Australopithecus* was under attack and, in a moment of panic, grabbed at a stone, which just happened to be chipped at the side, and swung it at the attacker. Surveying the damage caused by the blow, *Australopithecus* may have looked down at his new tool feeling relieved and impressed. Our observant primate might have returned home from the encounter and communicated with his kin about how effective the tool had been.

Alternatively, *Australopithecus* might have been working or even playing in the tribal boundary when two stones clashed against each other and a sharp edge was created. The experimental primate might have wondered, "What could I do with this?" The discovery would need to be explored and, upon finding that it cut effectively, he would begin to trial the tool.

Or, depending on how developed the *Australopithecus* brain was at anticipatory purpose, perhaps a pensive blue-sky thinker might have worked as a problem-solver. He might have observed several casualties from the tribe or an inability to extract all the protein from a kill, and decided to act to resolve the issue. He would have needed something that caused damage or was very sharp, and consequently the search was set in motion for a solution in the lands around them or for a way of manufacturing such a tool.

I wager that, either reporting back from the field, or from trialling or problem-solving within the camp, there would have been a gathering of interested brothers, sisters and elders to communicate, however primitively, about the tool's potential use. From this point on, *Australopithecus* might have begun to adopt a similar tool as a safeguarding weapon when hunting. Perhaps a few primates would have stayed back at base and further honed the tool so that it was handier, sharper, heavier, lighter or more attachable, depending on the need.

Curiosity   Case Observation

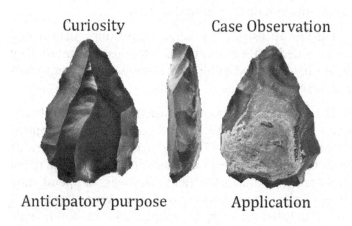

Anticipatory purpose   Application

Figure 35. Stone tool development would ideally have benefited from curiosity, case observations, anticipatory purpose and application.

Could this have happened if they had just been blue-sky thinking all the time? No. They needed a purpose to their thinking, such as reducing tribal casualties.

Could this have happened if they had just gone from one observation to the next? No. They needed to be iterating and refining their output so that it became better and better.

Could this have happened if they had simply applied the same ideas and technologies all the time? No. They would have needed to report back from successes, failures and circumstances; to share with others the challenges they had faced.

Could this have happened if there had been no feedback about how effective the bespoke, refined tools had been that day? No. They would have needed to continually refine and further develop their technologies to benefit even further from the discovery.

For our ancestors to achieve great things, they would have needed to have all of these functions. Therefore, they would have needed different people with different skills in different roles.

I believe altruism in applied science is analogous to the stone tool example. Albeit not exclusively, the applied scientist is the blue-sky thinking primate *and* the curious primate. Our work develops ideas from within the tribal boundary *and* out in the field. We need to spend time observing or contributing to the surveys, hunts and expeditions to fully grasp the demand and need. When we have acquired this experiential learning, we further develop these ideas with an intense focus on the utility of their output for the collective 'inclusive fitness' of the tribe.

Our ancestral developers would have needed to develop a rapport with the hunters and explorers. They would have needed to be trusted to develop relevant ideas. They would have needed to sign up to the big team goal of survival and tribal success.

They would have needed to be critical of their own ideas to be most successful. They would have needed to be adaptable and to think beyond what had been successful in the past, to ensure that the technology remained relevant and to continue to develop in response to changing demands. They would have needed to take accountable responsibility for the effectiveness of their ideas and to be able to get themselves into the minds of the hunters and their tutors. They would have needed to be constantly reflecting about their ideas and how they communicated the purpose and use of the tools, imparting the confidence necessary for the hunters and explorers to go out into the field knowing how to use their new methods.

To make an impact on the world, applied scientists cannot develop ideas and keep them to themselves, or even write the ideas up in a place where users cannot access them. If scientists develop their ideas or mechanistic insights without purpose, they are simply continuing down a line of investigation for their own ends. Scientists should be responsible for generating ideas in concert with the users. I wonder what the hunters, explorers and tribal elders, who truly appreciated the need to channel resources on survival, would have thought of the lone developer operating according to their own ideas without making them relevant to the big team goal and wider society.

Imagine that one of our ancestors made a stone tool, but was reluctant to show everyone or give guidance as to how a hunter might use it. Perhaps the developer would scratch an image of a stone tool on a faraway cave wall and expect the tribal members to go and find it in order to benefit from the tool. I believe developers need to aspire to be more altruistic by taking greater responsibility and ensuring that their work is designed to be more translational. No matter how fundamental, scientists should take every step possible to demonstrate the practical applications of their work to a societal need.

Altruism in applied sports science is infused with recognising the needs of the tribal bastions, fully immersed in the relevance of ideas and ways of ensuring the integration of ideas to the user. There is, of course, personal gain for applied developers. There is also gain for the hunters and explorers, and there is 'inclusive fitness' for the benefit of the tribe, but only when applied scientific developers work altruistically toward the tribal mission.

As a result, applied scientists should:

- Possess the humble and sensitive qualities of respectful service that a nurse displays, founded on an empathic understanding of the needs of others and gratified by the contentment of those they serve.

- Be committed to a goal in the same way the NASA cleaner was able to articulate his duties of being part of the bigger mission: sending a man to the moon.

- Be the Sherpas, with deep specialist knowledge of how to traverse, set up camp, cope with setbacks, explore new climbing techniques, lighten the load and advise the champion climbers, all in pursuit of the giddy heights of achieving the summit.

CHAPTER 9: HOW TO SUPPORT A CHAMPION: SUMMARY

"Life is a process of becoming, a combination of states we have to go through. Where people fail is that they wish to elect a state and remain in it. This is a kind of death."

Anaïs Nin

Throughout this book, I have focused on several 'flashbulb' moments in my career. No individual case is ever as simple as a single lesson, so I have focused on drawing out the big lessons and themes, throughout.

This chapter is a summary of those themes, some key observations and my top tips for developing yourself so you can increase your personal effectiveness in supporting a champion. Before I launch into that, here are a few other thoughts if you are aspiring to work in a similar world of nurturing people with an ambitious goal.

Before you support a champion...

If you are in a phase of 'learning' (A-levels, degree, postgraduate study or research, accreditation or certification or a personal development programme), there is every chance that the mode of learning you have experienced so far will be a very simple version of:

{Basic learning} = LEARN THEORY → WRITE-UP

I'm so sorry to break it to you, but this formula just isn't enough for the big bad world. Ideally, you would have an opportunity to experience the work environment that relates to your intended destination through a placement or volunteer work. This will enhance and translate your learning into a real-life scenario.

The icing on the cake would be if your course required you to not only learn about a topic, concept or theory, but to apply it to an actual person, population or function in a real-world setting before you process it, either by writing it up, discussing it or presenting it for your assessment. The formula for this type of work might look something like the following:

{Advanced learning method} =

LEARN THEORY → APPLY → SYNTHESISE

The 'synthesise' phase doesn't have to take the form of a multiple-choice exam or a written essay. What about presenting, pitching or debating? These practices are far more experiential.

Whatever your stage of learning, you can take some proactive steps to put yourself in the best position possible.

Firstly, club together with like-minded people to discuss, debate and critically question what you have read and learned. Repel the 'If it's published, it's fact' dogma.

Secondly, begin to advise others early in your studies. There is nothing quite like feeling the weight of responsibility of guiding others to sharpen your own approaches. When someone is looking to you to help them improve, it should intensify your questioning of the basic tenets, principles and knowledge concepts.

Finally, with unrelenting humility, patience and persistence, carve out an opportunity to influence a programme. It doesn't matter where your nearest sports clubs is, whether it's Telford & Wrekin Hockey Club, Manly Warringah Gymnastic Club, Spalding & District Indoor Bowls Club, Metz Race-Walking Club, the School of Pennsylvania Ballet, the Enniskillen Rehabilitation Unit or East Leake Triathlon Club, make the approach.

You will need to be hugely deferent to make the breakthrough of acceptance. Don't book a trumpet fanfare to celebrate your entrance: "I have knowledge from a book. I am therefore your saviour." Go along, knock on the door, and ask politely to speak to a coach or manager when they have a moment, not when they are busy. Tell them who you are and what you are studying, but, most importantly, ask if you can help with stopwatch timings, session setup, putting the mats out, getting the lane ropes organised, set up the noticeboards, do the photocopying or charge the batteries. It doesn't matter what you help with, just roll your sleeves up and get immersed in the microcosm.

While you're doing this, ask if you can learn about the programme. Find out why they are doing what they are doing, what goal they are working towards, using carefully chosen questions along the way. Only if they trust you will they ever turn around and ask you, "So, tell me about this stuff you're learning, have you actually read anything that real coaches can use?"

Then, with the preparation of a thousand hours of selective thought, reading, critique, observation, prioritisation and rehearsed pitching, you get to air your idea, suggestion or intervention. At this moment, you'll have become an applied practitioner. No longer languishing in just remembering the conclusion to an article, you are now an end-user of that knowledge; you are developing knowhow.

However, the process won't stop there. The coach or athlete might reject your idea. They might scoff at your best suggestion. This is when you will need to be able to reflect and react. Perhaps you will decide that this particular moment is not the right time. Maybe you didn't use the right words, or maybe your scrunched-up body language, with a rising intonation of self-doubt, suggested you weren't convinced either. You need to reflect, learn fast, adapt and set new standards for yourself. If you don't, you will get stuck at this level, as most do!

CHAPTER 1: TRUST

In a nutshell, trust is about developing rapport, establishing acceptance and gaining meaningful trust.

STAGE 1: DEVELOPING RAPPORT

One of the foundations of working with others and being a high-performing person is the possession of high self-awareness. Before you can take people with you, before you can influence, before you can form strong, meaningful and trusting relationships, you must know yourself. What is your style? How do you come across to others? What are you like on a bad day? What are you like when your strengths are overplayed?

My top tip would be to get feedback from people you trust implicitly. Asking people, "How did I get on today?" will normally lead to vague, positive praise. Asking direct questions, such as, "What could I have done to be more effective?" will give you a richer source of feedback. Impressions are formed in the first few seconds of any interaction, so work on this even if that means rehearsing what you're going to say.

Read: *Conversationally Speaking: Tested New Ways to Increase Your Personal and Social Effectiveness*, by Alan Garner[11].

STAGE 2: ESTABLISHING ACCEPTANCE

Once you are more self-aware, put your ego to one side and show people you're interested in them. Many people tell me they want to work with athletes, but they usually lead by sharing what they know. In this case, I would conclude that they are unlikely to be ready to support others. If, on the other hand, they lead with questions, they are more than likely ready. If they ask good questions, they will probably become competent practitioners. If, in a matter of a few questions, they can get to the heart of the issues, they will probably become outstanding practitioners. Questioning, listening and verbalising with another person about what you have heard shows that you have listened and acknowledged them and their thoughts.

My top tip would be to undertake a psychometric test and receive some objective feedback about your personality preference. If you can, use this to take your self-awareness to an advanced level, but also begin to recognise the preferences of other people and their personalities. This will enable you to flex your style and adjust your approach to better meet the needs of others. Remember, this is about adapting your style, not becoming someone else!

Remember, people don't care what you know, until they know that you care!

Read: *How to Win Friends and Influence People* by Dale Carnegie[12].

STAGE 3: GAINING MEANINGFUL TRUST

Acceptance should give you a solid foundation upon which to act. Trust is all about reliance and a confident expectation that something will be delivered. Trust takes time to build, but no time at all to erode. The support practitioner must go the extra mile to ensure that the basics are done very well.

My top tip would be that, if you are being asked to do too much, you need somebody to help you recalibrate your priorities. Gain clarity on what you are required to do, and then do everything in your power to do it to a level that will exceed expectations, but always focusing first on doing the basics brilliantly.

Read: *The Speed of Trust: The One Thing that Changes Everything* by Stephen M. R. Covey[13].

CHAPTER 2: TEAMWORKING

In a nutshell, teamwork is about taming your ego, putting ideas and pet projects to one side, signing up to a big team goal and working constructively with others on adding value to that goal.

STAGE 1: LET GO OF YOUR PET PROJECTS

We all have our special ideas and projects. They might have served you well in the past, sometimes extremely well. Perhaps they have enabled you to achieve good grades or to develop a specialism or a promotion.

As an applied practitioner, you must evaluate whether your idea is worth promoting. Ask yourself, would you take money from your own pocket and pay somebody to work on this project? If it doesn't pass this test, ask yourself how much it is worth pressing. There is a very clear difference between being paid to do a job in which you are promoting ideas that fascinate you, and being paid to do a job that gives a tangible return on performance.

My top tip for practitioners is to accept that they are specialist generalists. This might hurt some people's feelings because they want to have a unique specialism. But there is grandeur in being a generalist as your influence is wide-ranging. In a single consultation with a coach, you should expect to encounter a plethora of topics. You are the person who can influence in each of those areas.

Read: *A Short History of Nearly Everything* by Bill Bryson[14].

STAGE 2: SIGN UP TO THE BIG GOAL

When highly motivated people lack direction, they go off looking for the next big idea, often following their own interests. Time and again when the leader clearly states the big goal, this compels people to be part of that mission, and often they buy into it. People want to be inspired to work and a big goal is far more compelling than an isolated one. This is the power of collective purpose.

My top tip is to seek clarity from the leaders, the managers, the coaches, the directors. Ask them, "What is the big goal? What is the purpose?" If they're not sure, that might be concerning, but the question should stimulate them to develop their focus and then you will be influencing up. Super job! A further follow-up question would be, "What is the priority?" Often there are a myriad of factors to weigh up when deciding the priority, so ask for help about where you are putting your efforts.

Read: Russian folk tale, *The Little Red Hen*[15].

STAGE 3: CHANNEL YOUR RESOURCES

In a 24/7 society, an increasing number of demands will be placed on your time. For you to focus on the big goals, you will need to make room to immerse yourself fully in order to have the greatest impact. That means deprioritising everything else, which some might find sad or neglectful. Barack Obama famously asked his assistants to choose his suit for the day when he was US President. He prioritised the big decisions by removing the small ones.

My top tip is to apply some discipline to the way you work. Turn your email off for most of the day and attack it in one blast. Only look at social media in a set window (4.00pm to 6.00pm, GMT, is the hotspot of activity). That way, you should have more room in your day. Then go after that big goal, channel your resources and you'll be surprised by just how much you can achieve.

Read: *Essentialism: The Disciplined Pursuit of Less* by Greg McKeown[16].

CHAPTER 3: CURIOSITY AND CRITICAL THINKING

In a nutshell, pursuing curiosity and critical thinking is about leaving no stone unturned, questioning everything, while appreciating, acknowledging and harnessing the insight of the elite athlete and coach in exploring ideas.

STAGE 1: QUESTION EVERYTHING

The pool of knowledge is much revered, and right so in many respects. The record of scientific discovery will stand the test of time as a repository of our investigations, giving us an improved likelihood of outcomes. However, research has an estranged relationship with industry needs in tackling real-world problems.

For example, the concept of individualisation is critical in applied science. Published case studies are only just being recognised as useful work, up against a tsunami of randomised controlled trials and everything they represent.

My top tip, which comes with a heretical health warning, is to read the scientific literature differently. If a particular title, abstract and methodology has grabbed your attention, try looking beyond the average data. The variation in responses (standard deviation, for example) tells its own tale. Most people will dismiss a wide standard deviation of responses as a poorly controlled study or argue that, "Everyone is different. What do you expect?"

Instead, I see such a response as the fact that some are benefitting, whereas for others it needs to be unravelled. If scientists can ask the question: "How can we get the non-responder to respond and how can we get more response out of the responder?" they will be on a path to applied enlightenment!

Read broadly. Remember, we have embraced the idea of being a specialist generalist. Innovation will come from making new and diverse interconnections in your brain and between the brains within your team, not from burrowing down the same mineshaft and perpetuating a single doctrine! Get out and about, walk, run, swim and cycle. Brilliant ideas will arise when your mind is in a state of alpha waves[17].

Read: Anything by anyone that sparks your interest! Go on, follow your instincts!

STAGE 2: CREATE YOUR OWN EVIDENCE BASE

If the studies you draw upon involve specific observations or interventions, either from or with elite athletes, they might be more useful than the average piece of research. However, they are only a grain of sand compared with the diamond of collecting your own observations, and analysing and interpreting a dataset as it grows.

Sticking needles in athletes or asking them to fill in a sixty-five-point questionnaire on a daily basis is not possible, practical or fun. Scientific tools should be considered as tools and used to resolve performance questions rather than producing the outcome in itself.

My top tip would be for every practitioner to build a combination of objective, quantifiable observations along with a subjective, qualitative set of observations. These two sides of the coin complement your understanding to a much greater degree than either one in isolation. The very process of collecting such information will drive your own experimental design thinking and approach. It will drive your own anticipatory purpose, allow you to understand your athlete to a much deeper level and, ultimately, to give a much stronger answer to the coach.

Just as in fundamental science, you are dealing with probabilities, so it is better to know or to take that leap of faith from a strong evidence base rather than remaining in the dark and taking the leap based on blind optimism.

Read: *Better: A Surgeon's Notes on Performance* by Atul Gawande[18].

STAGE 3: LISTEN AND RESPECT

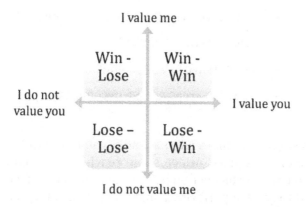

Figure 36. The 'Value me, value you' matrix. The aim of a successful relationship is to be in the top right-hand corner.

Referring to the above grid illustrates this point. You need to position yourself in the top right quadrant. At a bare minimum, you can exist in the bottom right, which might naturally occur during the early stages of a career or relationship as the Dunning-Kruger effect dips toward the nadir or as the rollercoaster lurches in any direction.

Awareness of your own strengths and weaknesses, and reflective practice, are too important to escape your attention.

My top tip is to question and listen to what athletes, coaches and colleagues have to say. If you find yourself stuck on the left-hand side, you might need to double your efforts. Are you caught in a downward spiral of criticism, bitching and always looking at the negatives? Everyone is different and it's likely that a host of factors underlie our behaviours. Dismiss people at your peril. Give all your respect to asking positive, neutral, open questions, such as:

- What is their natural perspective?
- Why have they come to that conclusion?
- What are they good at?
- What do we have in common?
- In which areas are we in complete agreement?

Questions like these will help shift you to the right-hand side of the square. Athletes and coaches are often more than happy to answer questions about how they feel, how well they have performed, and what their interpretation is. These case observations are gold dust in terms of developing your expertise and creating your own evidence base. The very nature of asking questions and seeking their opinions demonstrates that you respect them, their endeavours and their insights.

Read: *Just Listen: Discover the Secret to Getting Through to Absolutely Anyone* by Mark Goulston[19].

Chapter 4: Adaptability

In a nutshell, adaptability is recognising that talent comes in different shapes and sizes. Focusing on the main priorities allows practitioners to troubleshoot and take the necessary leaps of faith in applying science.

Stage 1: Accepting variety

Despite the same performance level, athletes come in different shapes and sizes, and they prepare in different ways at different rates with different outcomes. One of the most fascinating aspects of elite sport is trouble-shooting. This means identifying how an individual can improve without considering them clones of each other and rolling out the same advice time and time again.

My top tip is to shift from this fixed mindset to a growth mindset by celebrating the variety and seeing the holistic, unique individual as an exciting, bespoke, problem-solving challenge. Shake off the shackles of the randomised, controlled trial group mean, and embrace the standard deviation of the wildly different phenotypes, even at elite level.

Read: *Unweaving the Rainbow: Science, Delusion and the Appetite for Wonder* by Richard Dawkins[20].

Stage 2: Keeping things simple

If you computed all the relevant factors for an event, competition venue and individual athletes with their unique histories, and attempted to account for them all, you might never reach a performance-improving decision.

The number of variables in play is truly overwhelming. To move forward, you must treat your client's need like a good cooking sauce. You must reduce it.

My top tip would be to go as broad as possible, taking into account some of the most basic principles of life, such as:

- Stress overload and adaptation.
- Daily load and sleep.
- Consistency and variety.
- Slow and fast action.
- Depletion and repletion.
- Work and play.
- Building and peaking.

These and many other basic principles of vitality and life-balance should provide you with a simple decision-making framework for what to include in a programme. As General George Patton said: "A good plan today is better than a perfect one tomorrow."

You need to make decisions and move forward, no matter how complex the situation. Good athlete monitoring, note-keeping, reflections and constant review are guiding beacons to take with you as you move through any clouds of uncertainty.

Read: *Thinking, Fast and Slow* by Daniel Kahneman[21].

STAGE 3: TAKING A LEAP OF FAITH

All too often, you'll hear phrases like, "The limitation/trouble is…" In applied science, this type of talk kills people's imaginations. If you ever have the urge to stifle someone with, "The trouble is..." quickly signal to the substitute's bench, pull that tired player off the pitch and replace it with the fresh legs of, "But what if...?" It's effectively the difference between "No, but..." versus "Yes, and..." See what kind of difference it makes to you, those around you and those your champion.

Applied scientists have a much greater sense of the tension between existing knowledge from which useful knowledge can be plucked. They have a grasp on reality, variety, basic principles, and the deep skill and craft knowledge of working with others. They have the responsibility to bring knowledge to life.

Your crazy idea is ready to be born. It wants to wail its first scream and is hungry to be fed. Take a long run-up and enjoy the flight. You're about to take the leap of faith. Wahoo!

Albert Einstein expressed the following idea:

"If at first the idea is not absurd, then there is no hope for it."

Read: *Creativity: Flow and the Psychology of Discovery and Invention.* by Mihaly Csikszentmihalyi[22].

CHAPTER 5: ACCOUNTABILITY AND RESPONSIBILITY

In a nutshell, accountability and responsibility are about adopting and appreciating the mindset of the coach (or client) and the responsibility they take for results. Practitioners should acquire the same sense of accountability for their impact and influence.

STAGE 1: WALK IN SOMEONE ELSE'S SHOES

It can be all too easy to exist within the boundaries of the normal behaviours of a job description. It may be easy to hide behind your professional standing and think about what *you* are doing and how *you* are getting on.

My top tip is to devote time to thinking about who you hold key relationships with. Once you have identified these important folk, ask yourself what pressures they are under. What keeps them up at night? What are their needs? How would it feel to do what they do?

It would be even better if you could find a way of truly experiencing or assuming a similar role; either through shadowing or, better still, by getting involved yourself. Working with coaches requires empathy of the coaching process. What better way to empathise than to have walked the walk?

Read: *To Kill a Mockingbird* by Harper Lee[23].

STAGE 2: DOING IT WELL

The word 'artisan' refers to a person who is skilled in an applied art; a craftsperson or company that makes high-quality or distinctive products in small quantities, usually by hand or using traditional methods.

Taking full accountability for the responsibility of delivering something means you will need to attend to your craft of delivery to the highest order. Artisanal delivery in applied science requires you to nurture your products, but also to cherish the feedback of others and fully espouse the notion of improving your work for someone else's sake.

My top tip here is to adopt a standpoint. There are the trolls who claim that any type of science and medicine with athletes involves doping, but that's because they are bathed in ignorance and a simple lack of understanding. In this day and age, athletes have a viable alternative method of performance enhancement to doping. That alternative is applied science, which also happens to be compliant with the rulebook. Whatever triggers doping – losing, greed, poor ethics or desperation – could just as easily activate an athlete to reach out to a team of applied scientists to work on their performance. Oh, and it's so much better on the soul!

This mindset should lead you to focus on doing things well, offering artisanal delivery, promoting ethical science and championing clean sport.

Read: *The Secret Race: Inside the Hidden World of the Tour de France* by Tyler Hamilton and Daniel Coyle[24]. This book shows where a lack of compelling and ethically sound applied science can lead.

STAGE 3: TAKING ACCOUNTABILITY

The purity of sport is in its absolute outcome. People who enjoy sport recognise that losing is part of the character-building aspect of facing the outcome. Sport is wonderful and brutal in the same breath.

When you sign up to the big team goal, you sign up to the potential for success, and you do so as a collective. If the multidisciplinary team members point the finger of blame when poor performance occurs, it is tantamount to having a hissy fit at the end of a school sports day race!

My top tip is to wear your accountability on your sleeve. It will act as a shield and guiding light to your actions. Embed it into your day-to-day mantra and it will develop a unified and sharing philosophy among others. Standing up and being counted is not only progressive; it is the ultimate symbol of teamworking.

Read: *Will it Make the Boat Go Faster?: Olympic-winning Strategies for Everyday Success* by Ben Hunt-Davis and Harriet Beveridge[3].

CHAPTER 6: INFLUENCE

In a nutshell, influence is recognising that words and actions can have a powerfully negative or positive effect on the performer. Therefore, great care is required.

STAGE 1: BE AWARE OF YOUR THOUGHTS

A disorganised mind can be a bugger to wrestle with. Tuning into the good thoughts and telling the bad ones to shut up can be a constant internal struggle. Intrapersonal strength is required in applied practice because not everything you think is helpful to the unique circumstances in that moment in time, and with that specific person.

My top tip is to devoutly and diligently undertake reflective practice. It will strengthen and mature you, and it will bring balance to the force.

Read: *A-Z of Reflective Practice* by Fiona Timmins[25].

STAGE 2: SHARE YOUR THOUGHTS

Your thoughts and reflections should be well-honed before they leave your mouth. You now need to find someone you can trust to share your deepest, darkest, craziest crackpot ideas with. If you haven't already accumulated some trusted colleagues and friends throughout your career, it's well worth jumping into a time machine and going back to find some along the journey!

My top tip is, once you've done your own evaluations, share your ideas with your close confidants first. Talk with your mentor if you have one. The very exercise of sharing an idea will take you halfway to realising whether or not it's a good one. Trusted friends and colleagues should be considered as such if they give you an honest evaluation of your thoughts and ideas. Use them to pilot your thoughts; not the coach or athlete. This is a safer approach and will mature your idea before it reaches the stratosphere of your client's earhole.

Read: *Thanks for the Feedback: The Science and Art of Receiving Feedback Well* by Douglas Stone and Sheila Heen[26].

STAGE 3: PLAN THE HEIST

Determined, motivated and interested people will have good ideas. Occasionally, they will have great ideas. Every now and again, those ideas will get the green light, be given some rocket fuel and blast into orbital brilliance. The occasional great idea will address the big issue; the so-called, 'elephant in the room'. Launching, unsolicited, into probing questions, as I did with Matt Pinsent, isn't to everybody's taste.

My top tip for dealing with this is to pull a plan together equivalent to organising a 'heist'. This idea comes to you courtesy of brilliant psychologist Dr Mark Bawden.

It takes patient persistence, so take a deep breath in and slowly exhale. First, you need to identify the problem. This means stripping the issue back completely, freeing it of any entanglements, your emotions, other people's emotions and all the circumstances from which a bonfire of excuses can ignite. Ask yourself the following questions:

- What is it I want to affect?

- Who is it I want to influence? When are they most receptive? When are they least receptive?

- How does the target[xxx] like to receive information? Does he or she like to think or talk? When they talk, do they prefer to do so on a one-to-one basis or in groups? Do they like to analyse data and take their time to digest the problem, or do they like the summary headlines and making impulsive decisions? Is there someone else who is more influential than you who could sow the seed of the idea?

- What are their interests outside work and how much do you discuss these with them?

- How have other ideas been accepted by your target? Where was it proposed? What were they doing at the time?

Here are some determining criteria for a hypothetical heist:

- **Target:** Coach A: female, mid-forties, former international athlete.

- **Personality observations:** Fiery, comfortable with decision-making, dynamic, charismatic, focused.

- **Least receptive:** During the middle of a training session.

- **Most receptive:** On a plane going to a training camp, or in the stands immediately after a World Championships or Olympics.

- **Previous ideas have been successful when:** She is on her own, rather than in a group discussion; she was sitting on the balcony of a

---

[xxx] *Oh yes, we are doing this properly – 'the target'!*

restaurant overlooking an idyllic training venue in Switzerland.

- **Outside interests:** Dogs, skiing and wine.

If you were to consider all these factors together, I expect your approach to the launch of your idea would be very different. Hopefully, it would dissolve any impatience to deal with it immediately, in an off-the-cuff or underprepared way. Your sense of urgency should be overwhelmed by your sense of executing successfully. Ideas are transitory and conditional, so find the right moment with the right conditions and pitch it well. Adopt the patient and persistent precision of a sniper.

Read: *Agent Zig-Zag: The True Wartime Story of Eddie Chapman: Lover, Traitor, Hero, Spy* by Ben McIntyre[27].

CHAPTER 7: HANDLING UPS AND DOWNS

In a nutshell, high-performance sport is full of ups and downs. Practitioners need to know themselves, understand their values, be up for a challenge and be prepared for the turbulence.

STAGE 1: VALUE YOUR VALUES

No one can shake you from your values. They are the representation of you in innate principle form. When the waters get choppy, your values will keep you from getting seasick. When the wind gets up, your values will give you wings. To balance and handle the highs and lows of the rollercoaster, your values are the bedrock reference point from which you will obtain comfort, solace and strength.

My top tip is to do a 'values decision' exercise. For example, what are your top nine values? Write a list of nouns you associate with yourself. Below is an example list to get you started.

Once you have your values, share them with someone else. Talk through why they are important to you so they are cemented in your mind.

Read: *Long Walk to Freedom* by Nelson Mandela[28].

Equality	Forgiveness	Courage	Trust
Wealth/ prosperity	Friendship	Loyalty	Ambitious
Patience	Self-respect	Curiosity	Challenge
Harmony	Competition	Empathy	Communication
Productivity	Intelligence	Innovation	Pleasure
Discipline	Responsible	Social recognition	Honesty/ Integrity
Security	Independence	Power	Teamwork
Decisiveness	Simplicity	Politeness	Happiness
Excitement	Helpful	Risk-taking	Effectiveness
Family	Spirituality/ faith	Freedom	Quality
Health	Respect	Variety	Achievement
Open- mindedness	Hope	Growth	Humour
Creativity	Strength	Service	Beauty

Figure 37. Table of values.

STAGE 2: AUTHENTICITY

Life, in all its glory, will unveil new experiences, new people, new ideas and new realisations for you to absorb or ignore, but ultimately you are stuck with you. You will have to spend your whole life putting up with yourself;

there's no getting away from that. We have already aspired to develop self-awareness, but if you want to develop deep and authentic personal leadership, awareness isn't enough.

Authentic leadership is a progressive modern concept. It is obviously focused on leading others, either by influence or direct responsibility, but it also speaks to leadership of self. Authentic self-leadership should be an aspiration for all; independent of stage and age, status and standing.

My top tip to achieving authenticity is to take reflective practice to a new level and share your lessons and reflections with others.

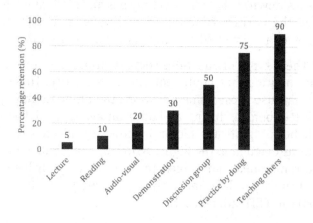

Figure 38. Potency of learning styles and methods (attributed to National Training Laboratories, Bethel, Maine).

The potency of learning styles and methods above provides us with a useful reflection of how powerful sharing your experiences is in developing your own learning and approaches. Pulling your experiences, ideas and interpretations together matures your personal leadership far more than just reflecting on your own ideas in a darkened room. Get out there and tell your story. You will discover what you believe in to a whole new depth.

Read: *True North: Discover your Authentic Leadership* by Bill George and Peter Sims[29].

STAGE 3: PLAN FOR THE HIGHS AND LOWS

My top tip is to frequently ask yourself: 'What's about to occur and how am I going to feel about it?' A tried and tested exercise for Olympic managers, coaches, athletes and support staff is to ask, "What if…? What if the bus breaks down? What if I forget my passport? What if I don't have the correct running spikes? What if the stadium showers are cold? What if the plane food is awful?"

These types of questions serve to prepare for the many eventualities that could potentially transpire. However, you can use these same questions to anticipate how you will feel before, during and after an event. Your feelings will be a significant factor in how any given situation manifests itself. In the highs, switch on your bullshit filter and remember your values. In the lows, fill your life with the things you love more than you love yourself (taken from the TED talk below).

Watch: 'Success, Failure and the Drive to Keep Creating' (TED talk) by Elizabeth Gilbert[30].

CHAPTER 8: ALTRUISM

In a nutshell, altruism means that the world is a better place if you are in it to support, care and develop, rather than to grab the limelight.

STAGE 1: ACCEPT YOU HAVE AN EGO

I have had the privilege (and it is an utter, utter privilege) of working with more than a thousand athletes. Of that group, more than two hundred have gone on to achieve a medal at the World Championships or Olympic Games. Many would have won regardless of the type of input I provided, and some probably would have won without my input.

Many medallists and non-medallists have thanked me for my help, but I don't do the job I do for the thanks. I do it for the sense of purpose I get from developing others! However, thanks and congratulations will give anyone's ego a warm hug. Occasionally, it will feed you with a spoonful of narcissism, so be careful what you do with it.

My top tip is no more complicated than accepting that you have an ego, and that it needs looking after. Balance should be sought.

Read: *The Blank Slate: The Modern Denial of Human Nature* by Steven Pinker[31].

STAGE 2: THE JOY OF SERVICE

Giving is better than receiving. At least, we try to convince ourselves of this when it comes to Christmas. To me, supporting others is equivalent to the highest peaks in the

Himalayas, whereas my own achievements amount to the foothills in comparison. However, seeing people I have worked with reach the highest sporting achievements is no different from having worked in a post-cardiac rehabilitation unit and a patient announcing to the staff, "Your help has changed my life." What greater feeling could anyone have than in helping others?

Read: *Authentic Happiness: Using the New Positive Psychology to Realise Your Potential for Lasting Fulfilment* by Martin Seligman[32].

STAGE 3: NORMALLY CONFLICTED

If you have given your ego a break and given your humility a companion, and if you have then fallen in love with working in service to a bigger goal than your own, you will still need to introduce rewards and motivating factors. At times, you'll feel the conflict of why you do what you do. Every now and again, existential questions will arise and might even torment you. This is normal. Giving is nice but so is receiving, and there will be lows after the highs.

My top tip is to practise mindfulness. This simple technique asks you to tune in to your stream of consciousness. Mindfulness encourages you to pay attention to what is in your mind, and the purpose of it being in your mind, in a non-judgemental way. Raising your awareness and acknowledging your thoughts will give you clarity and greater acceptance of your reality.

Read: *Mindfulness: A Practical Guide to Finding Peace in a Frantic World* by Mark Williams and Danny Penman[33].

If you do only half of the above, you'll be in great shape to support a champion.

Chapter 10: A Final Word from Me

*Champion [**cham**-pee-uh-n]*

NOUN: A person who fights or argues for a cause or on behalf of someone else.

The Oxford English Dictionary

All the lessons shared in this book took seed from my ambition to work with the best athletes in the world. I was fuelled by my own wonder of watching sport throughout my childhood, most notably the 1992 Barcelona Olympics, when I gazed in awe at my heroes trying to win and sometimes doing so.

I was spurred on by inspirational mentors such as Colin Clegg throughout my A-level studies, given a break by Dr Andy Miles, tutored by Peter Keen, Prof Jo Doust and Dr Helen Carter throughout my studies and by Dr Ken van Someren throughout my career. They showed me why people spend their lives learning, questioning and driving themselves forward to achieve meaningful goals.

I was either lucky or shrewd to develop a bank of experiences before I encountered 'real' athletes. Working with Redgrave and other top athletes from the summer of 1998 required me to understand the world from their point of view. This might seem impossible when you consider that their talents are so extreme, so alien to us mere mortals, and so far removed from our day-to-day lives. But these elite athletes are still human. They have the same thoughts, insecurities, hopes, dreams and ambitions as the rest of us. What separates them is that they have physical gifts *and* the will to act upon them.

If my relationships with athletes have blossomed, it is because the empathy I have shown, and my willingness to work to their goals, has exceeded my own role and standing on the team. The work of support staff is indeed paradoxically ambiguous. Altruists who work in sport will often give unconditionally, but will usually also make personal gains. Therefore, they are likely to be conflicted and will probably question at times what the hell they're doing working with people who run, jump, chuck stuff, crank wheeled vehicles, kick leather balls or splash about for a living.

However, sport continues to be something that the human species desires. It uplifts us, gives us goosebumps, and develops a real sense of pride and worth for the masses. So the world of sport and talent needs people who are prepared to support our champions; to be the team supporting the team. Applied science is here to stay in sport. We are here to help athletes and coaches test the boundaries of what our tribal bastions can legitimately achieve.

For anyone who works with or wants to work with elite athletes – and for anyone who works with any other type of talent, be it a prima ballerina, a virtuoso violinist, a Cirque du Soleil contortionist, an astronaut, a military commander, a surgeon, a business leader, a playgroup leader or anyone who has the opportunity to unlock talent in others – I would say the following.

Just knowing something isn't enough. Applying what you know through learning about who you are and respecting who you work with; being honest with yourself; signing up to a significant challenge; and developing new and innovative ways to problem-solve together as a team will bring your knowledge to life. It will help others and will more than likely give your life greater meaning and contentment.

I hope this book helps you find a better way of supporting whoever or whatever you champion.

CHAPTER 11: A FINAL WORD FROM THE ATHLETES AND COACHES

"Champions are not the ones who always win races, champions are the ones who get out there and try. And try harder the next time. And even harder the next time. Champion is a state of mind. They are devoted. They compete to best themselves as much if not more than they compete to best others. Champions are not just athletes."

Simon Sinek

SIR STEVE REDGRAVE, CBE, MBE

"Five Olympic gold medals do not come easy. An athlete requires unremitting focus and the willingness to explore all possibilities for self-improvement. Support staff need to adopt the same approach in developing their own performance. *How to Support a Champion* is a good read for all those working in high performance."

Martin McElroy

"The key to being a successful applied sports scientist is the ability to recognise the key areas where the coaches both want and need support. The old saying that 'If the only tool you have is a hammer, then everything starts to look like a nail' is particularly apt in sports science. The challenge is not to show how smart the scientist is. The challenge is to be a performance partner in the team.

"Coaches are applied practitioners. Sports scientists must be applied scientists, helping the coach identify and develop workable performance solutions. Working as a coach with Steve in the run-up to the 2000 Games in Sydney, I had a real performance partner who could identify with the challenges and come up with real working solutions."

Hayley Tullett

"My coach Mark and I were always searching for answers. Most of the scientists just turned out numbers and didn't make them useful for us. When we first started working with Steve, we drove him hard to make the science more and more relevant to us. I'm curious and suspicious by nature, so he needed to prove why the theory should work in practice before I put it to the test.

"Being forward thinking, Steve was up for the challenge and regularly presented his new training, and recovery theories and insights. Importantly, over time, we could see how he was adapting and applying the knowledge and theories so it was tailored to our individual makeup. This personalised application was what I think triggered some of our biggest gains!"

MARK ROWLAND

"When I first got in contact with Steve in 2002, I knew he had a good background in middle-distance events and had worked with several good runners. But what I needed was someone I could trust and talk to; a scientist who wasn't just reliant on studies or kneejerk advice, but one who was able to really question what I was doing, what my athletes were doing and be able to really individualise their advice to our needs. In a nutshell, I needed to challenge what I was applying and to know if Steve was talking bollocks!

"Working together with Steve, he was able to research new ideas, adapt and develop, but so was I. So together we were better able to support my athletes.

"To be able to really get the most out of coaching and science, both fields have got to work in unison for the benefit of the athlete."

DAME JESSICA ENNIS-HILL, CBE

"When I am stood on the start line, I need to know I have prepared in a meticulous way. Using applied science gives me confidence to perform to my best. Guesswork and opinions alone leave too much to chance. Simply transplanting scientific studies into my day-to-day training would be madness.

"Throughout my career, Steve has provided me with a way of making science, objectivity, innovation and ideas useful for my preparation and performance. If science can be applied to the complex world of heptathlon, it can probably be applied to anything. This book shows you how."

Toni Minichiello

"Scientists can use all the big words they like, but if they can't actually make their knowledge and methods useful, then it's no use to me. I don't need people to come in and be all 'flash Harry', I need people to work with me, become part of the team and become part of the family working together to improve performance.

"Steve was given the challenge of working with the heptathlon and thinking imaginatively through the details and the philosophy; the science and the emotions of performance. Steve's help continues to be invaluable to help Jess perform at the right moment to a level we couldn't have imagined back in 2004.

"This book is invaluable for anyone working not only with talent, but for anyone working with other people towards a big goal."

Kelly Sotherton

"When you exhaust the conventional beliefs in your sport, the search for something different is hard. Taking a chance on working with a scientist who has no evidence base for your specialist area is a big step. Firstly, you have to believe and trust in your new journey. This was my big step in 2007 when I needed to move up another level.

"Even with other athletics coaches stating that this was wrong for me, we persisted. It paid off. A great example was in 2008 when I missed half the winter, had only six weeks of running and then produced 60m hurdles, 400m and 800m personal bests indoors.

"Doing what I had always done was not an option, but change is an unnerving path. Steve showed me how science can be applied in a creative way, experimenting to find a new way of getting the best from me. I showed him how he could do so and be accountable for my results.

"I know he works differently as a scientist and a leader as a consequence, and now, in my coaching, applied science has left an indelible mark on how I work with athletes and in business."

SIR MATTHEW PINSENT, CBE, MBE

"There are very few people that I would consider an insider to our efforts to win at successive Olympics, but Steve was one of them: utterly professional, but with a lightness of touch and a sense of humour that made him instantly likeable and one to trust. He was key to our challenge in 2001 to win two World Championship golds in two hours. We wouldn't have achieved it without him.

"*How to Support a Champion* tells it like it is, and is a superb insight into what is required of anyone who supports other people, let alone elite athletes."

JAMES CRACKNELL, OBE

"Throughout the early stages of my career, especially the 1996 Olympic Games, where I was forced to withdraw on the eve of the Opening Ceremony, I was frustrated with not only the inconsistency of my performances, but believed my optimum could and should be far higher. I am sure most athletes think this way.

"From 1996 onwards, I saw a real step up in my performances due to a professionalisation of my approach and a big part of that was the application of science to my training and competition preparation.

"It's one thing knowing all of the things you need to do, but you also need a partner to help you prioritise and incorporate the ideas into your daily schedule. Steve helped me do this. I saw the benefits of his analytical but personal approach, brutally constructive honesty and his commitment to helping me go faster."

WORKS CITED

*A good book should leave you…slightly exhausted at the end.
You live several lives while reading it."*

William Styron

1. Kruger, J. & Dunning, D. Unskilled and unaware of it: How difficulties in recognizing one's own incompetence lead to inflated self-assessments. *J. Pers. Soc. Psychol.* **77**, 1121–1134 (1999).
2. Gibbs, G. *Learning by Doing: A guide to teaching and learning methods.* (Further Education Unit. Oxford Polytechnic, 1988).
3. Hunt-Davis, B. & Beveridge, H. *Will it make the boat go faster? : Olympic-winning strategies for everyday success.* (Troubador Pub., 2011).
4. Clegg, C. *Exercise physiology.* (Feltham Press, 1995).
5. Ingham, S. A. *et al.* Determinants of 800-m and 1500-m Running Performance Using Allometric Models. *Med. Sci. Sports Exerc.* **40**, 345–350 (2008).
6. Ingham, S. A., Fudge, B. W. & Pringle, J. S. Training Distribution, Physiological Profile, and Performance for a Male International 1500-m Runner. *Int. J. Sports Physiol. Perform.* **7**, 193–195 (2012).
7. Krakauer, J. & Rackliff, R. *Into thin air : a personal account of the Mount Everest disaster.* (Anchor Books, 1999).
8. Rees, T. *et al.* The Great British Medalists Project: A

Review of Current Knowledge on the Development of the World's Best Sporting Talent. *Sport. Med.* **46,** 1041–1058 (2016).

9. Hamilton, W. D. The genetical evolution of social behaviour. II. *J. Theor. Biol.* **7,** 17–52 (1964).

10. Montesinos, J. *et al.* Barcelona baby boom: does sporting success affect birth rate? *BMJ* **347,** (2013).

11. Garner, A. *Conversationally speaking.* (McGraw-Hill, 1981).

12. Carnegie, D. *How to win friends and influence people.* (Vermilion, 2006).

13. Covey, S. M. R. *The speed of trust : why trust is the ultimate determinate of success or failure in your relationships, career and life.* (Simon & Schuster, 2006).

14. Bryson, B. *A short history of nearly everything.* (Black Swan, 2003).

15. Randall, R. & Pichon, L. *The little red hen.* (Ladybird, 2012).

16. McKeown, G. *Essentialism : the disciplined pursuit of less.* (Virgin Books, 2014).

17. Lagopoulos, J. *et al.* Increased Theta and Alpha EEG Activity During Nondirective Meditation. *J. Altern. Complement. Med.* **15,** 1187–1192 (2009).

18. Gawande, A. *Better : a surgeon's notes on performance.* (Metropolitan, 2007).

19. Goulston, M. *Just listen : discover the secret to getting through to absolutely anyone.* (American Management Association, 2010).

20. Dawkins, R. *Unweaving the rainbow : science, delusion, and the appetite for wonder.* (Penguin, 2006).

21. Kahneman, D. *Thinking, fast and slow.* (Penguin, 2011).

22. Csikszentmihalyi, M. *Creativity : the psychology of discovery and invention.* (Harper Perennial, 1997).

23. Lee, H. *To kill a mockingbird : Harper Lee.* (Harper Perennial, 2006).

24. Hamilton, T. & Coyle, D. *The secret race : inside the hidden world of the Tour de France : doping, cover-ups, and winning at all costs.* (Bantam Books, 2012).

25. Timmins, F. *A-Z of reflective practice.* (Macmillan Education UK, 2015).

26. Stone, D. & Heen, S. *Thanks for the feedback : the science and art of receiving feedback well.* (Viking, 2014).

27.   Macintyre, B. *Agent Zigzag : the true wartime story of Eddie Chapman : lover, betrayer, hero, spy.* (Bloomsbury, 2007).

28.   Mandela, N. *Long walk to freedom : the autobiography of Nelson Mandela.* (Abacus, 1994).

29.   George, B. & Sims, P. *True north : discover your authentic leadership.* (Jossey-Bass, 2007).

30.   Gilbert, E. No Title. Available at: https://www.ted.com/talks/elizabeth_gilbert_success_failure_and_the_drive_to_keep_creating.

31.   Pinker, S. *The blank slate : the modern denial of human nature.* (Viking, 2002).

32.   Seligman, M. E. P. *Authentic happiness : using the new positive psychology to realize your potential for lasting fulfillment.* (Free Press, 2002).

33.   Williams, M.,& Penman, D. *Mindfulness: a practical guide to finding peace in a frantic world.* (Hachette, 2011).

34.   Winter, E. M. *et al. Misuse of 'Power' and other mechanical terms in Sport and Exercise Science Research. Journal of Strength and Conditioning Research* **30,** (2015).

Thank you for reading.

Now get out there and support a champion!

Steve

About the Author

Dr Steve Ingham is one of world's leading performance scientists. A physiologist by trade, he has a track record of providing scientific support to more than a thousand athletes, of which more than two hundred have gone on to achieve World or Olympic medal success. Steve has also coached athletics to World and Olympic medals.

Steve holds a BSc from the University of Brighton and a PhD from the University of Surrey. He is a fellow of the British Association of Sport and Exercise Sciences.

Steve was previously Sports Science Manager at the British Olympic Association, and the Head of Physiology, then Director of Science and Technical Development at the English Institute of Sport.

He now runs Supporting Champions dedicated to providing transformational personal development in sport and business around the world.

Steve is a popular motivational speaker on the topics of: lessons from high performance, developing high performance teams and developing your model of performance.

MORE FROM STEVE INGHAM

Twitter: @ingham_steve

Blog: www.supportingchampions.co.uk/scblog

www.supportingchampions.co.uk

NOTES

Printed in May 2022
by Rotomail Italia S.p.A., Vignate (MI) - Italy